Systematic
Geography of Jammu and Kashmir

Systematic

Geography of Jammu & Kashmir

Majid Husain

RAWAT PUBLICATIONS

Jaipur • New Delhi • Bangalore • Guwahati • Kolkata

ISBN 978-81-7033-571-9 (paperback)
ISBN 978-81-7033-572-6 (hardback)

Reprinted, 2022

Published by
Prem Rawat for **Rawat Publications**
Satyam Apts., Sector 3, Jawahar Nagar, Jaipur - 302 004 (India)
Phones: 0141 265 1748 / 265 7006
E-mail : info@rawatbooks.com
Website: rawatbooks.com

New Delhi Office
4858/24, Ansari Road, Daryaganj, New Delhi 110 002
Phone: 011-23263290

Also at *Bangalore, Guwahati* and *Kolkata*

Typeset by Rawat Computers, Jaipur
Printed at Nice Printing Press, New Delhi

To the new generation
of the 21st Century

Contents

Preface

The study of earth's surface as the space within which the human population lives is an important branch of knowledge. It deals with the living and non-living things and their interrelationship found on the earth surface. The present book '*Systematic Geography of Jammu and Kashmir*' is a higher level textbook, designed and written to cater the needs of under-graduate and postgraduate students of geography. Based on twelve chapters and documented over sixty original maps and diagrams, the book systematically explains the physical personality and cultural diversity of the northern most state of the country. Apart from perusing the available literature and data on the state, the author utilized his first-hand experience in the book while staying in the state for about a decade in the capacity of Professor and Head of the Department of Geography, University of Kashmir, Srinagar.

The book begins by presenting a concise account of the structure, relief, drainage pattern and major lakes of the state. The climatic elements, i.e., temperature, humidity, precipitation, and the way they have been influenced by factors such as changes in latitude, altitude and topographical features have been examined in the second chapter. The natural vegetation climax and the spatial distribution of soils and their textures and structures have been discussed in the third and the fourth chapters respectively, while the subsequent chapter deals with the mineral resources of the state.

The people of Jammu, Kashmir and Ladakh divisions have different ethnicity and their culture is usually different from each other. The cultural mosaic and demographic attributes of the three administrative units of the state have been succinctly presented in the sixth and the seventh chapters. The state of Jammu and Kashmir is essentially agrarian in character and its basic economic activities have been closely influenced by the physical environmental factors. This point has been underscored in the chapter on agriculture in which the agricultural landuse, cropping intensity, crop combination and agricultural productivity regions have been discussed. Despite abundance of water, forest and some of the valuable minerals, the state is lagging behind in industrial development. There are some industries like carpet making, woollen textile, paper mache, agro-forest based industries, etc., which are being carried on in the state right from the ancient period. These industries are however, facing tough competition from the outside world. The industrial scenario of the state and the problems associated with the industries have been examined in the ninth chapter. Tourism is an emerging sector in the economy of the state. It contributes about one-fifth to the state income and provides employment, directly and indirectly to thousands of workers. Tourism, thus, has the potential for bringing major economic development and social transformation. Tourist activity in the state, however, suffers from problems of seasonality and creating the problems of cultural pollution and ecological degradation. The tourism potential and its socio-economic and ecological implications have been discussed at length in the tenth chapter. The processes and levels of urbanization in the state have been elaborated in the eleventh chapter, while the last chapter gives a brief account of the means of transportation and communication and the salient characteristics of trades, imports and exports in the state.

My colleagues and students in the Department of Geography, University of Kashmir, where I spent the best part of my teaching career from 1980 to 1987, have contributed in various ways in the preparation of manuscript of this book for which I feel highly indebted. The author is grateful to Mr. K. Rawat, his sons, family and Rawat Publications, Jaipur for their encouragement in the completion of this project. All the maps and diagrams of the book were neatly drawn by Mr. Ganga Sharan applying his cartographic skill for which he deserves special thanks. Dr. Haseena Hashia, Mr. Javed Alam, Dr. Shagufta Bano Husain, Abdul-Rahman Mahdi, Mansoor Majid provided me moral support to accomplish this task for which I feel thankful. The inquisitive nature and innocent behaviour of my grand-children Majid Abdul Rahman, Madeeha Javed, Jibran Mansoor and Rashad Abdul Rahman were quite refreshing and stimulating. I thank them and wish them long, happy and prosperous life. I also wish to thank my wife who helped me in more than one way in the preparation of the book. All the views, commissions and omissions in the text are entirely the responsibility of the author. The author would be appreciative of the constructive criticism aiming at improvement of the subsequent editions of the book.

Majid Husain

1

Structure and Relief

The scientific study of rock structures, including their form, genesis and spatial distribution, but not their compositions is known as structure in geology. The state of Jammu and Kashmir is the finest repository of sequence of rocks right from the Archaean Era to the Recent Age (present day alluvium, etc.). The present chapter gives a concise account of the rock structures and their spatial distribution in the state of Jammu and Kashmir.

Archaean (Archean) System

Archaean is regarded as synonymous with Precambrian, but sometimes used to refer to older Precambrian rocks. The oldest rocks of the earth's crust that have been found at the bottom of the stratified deposits are known as Archaean rocks. They form the core of all the great mountain chains of the world and the foundations of all the great ancient plateaus. They are all *azoic*, thoroughly crystalline, extremely contorted, faulted, largely intruded by plutonic intrusions, and generally have a well-defined foliated structure. Owing to their extreme complexity of character, the Archaean System is also known as the "*Fundamental Complex*", or the "*Basement Complex*" (Fig. 1.1).

The origin of Archaean rocks is a point of controversy among the geologists. Various modes of formation have however, been ascribed to these rocks. In the opinion of some experts, they represent, the first formed crust of the crust by the consolidation of the gaseous or molten planet. Some are believed to be the earliest sediments formed under conditions of the atmosphere and of the oceans in many respects different from those existing at later dates, and afterward subjected to an extreme degree of thermal and regional metamorphism. Some are believed to be the results of the consolidation of the original heterogeneous magma erupted excessively in the crust, while some are thought to be the result of the bodily deformation or metamorphism of large plutonic igneous masses under great earth movement or stresses[1].

FIGURE 1.1

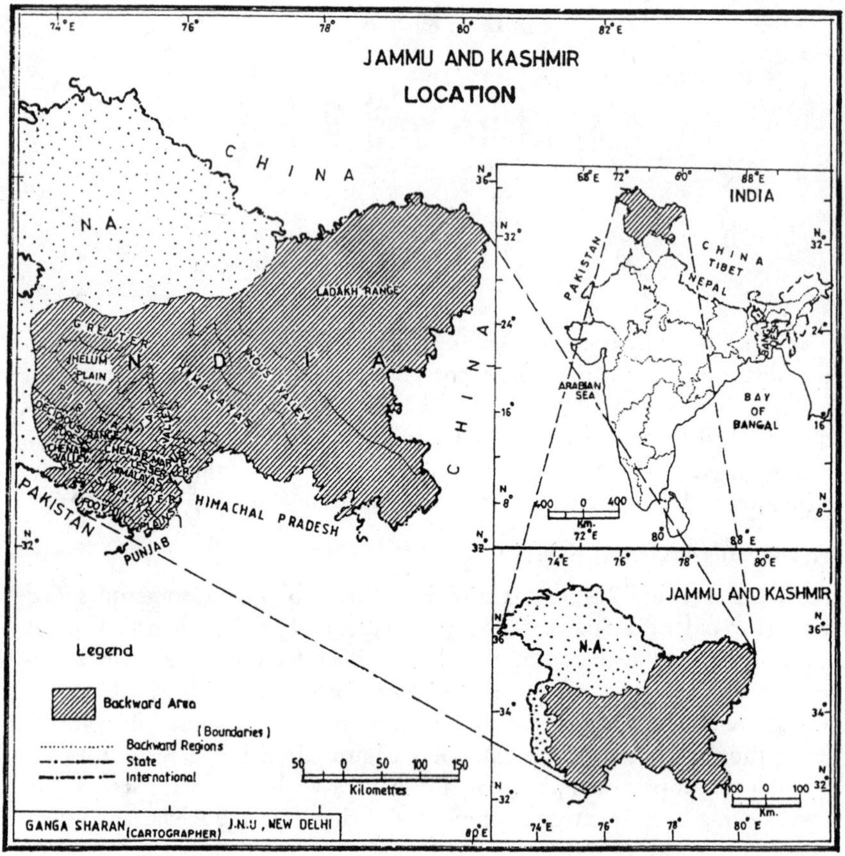

The metamorphosed rocks of the Archaean System are widely spread over the Peninsular India, Chotanagpur Plateau and Bundelkhand. They are also found along the whole length of the main Himalayas, forming the bulk of the high ranges and the backbone of the mountain system.

The most common Archaean rock is gneiss—a rock which in mineral composition may vary from granite to gabbro, but which possesses a constant, more or less foliated or banded structure, designated as gneissic. The bulk of the high ranges of the Himalayas forming the Central or Himalayan zone is formed of crystalline or metamorphic rocks, like granites, granulates, gneisses, phyllites and schists. The high snow-peaks of the central axis have a substratum of these rocks.

The North-western Himalayas of Kashmir to the north of the Middle

Ranges, forming the core of the Dhauladhar, the Zanskar and the ranges beyond in Ladakh and Baltistan are occupied by gneisses and schists. These rocks were all regarded as igneous and called 'central gneiss' by Stoliczka and were taken to be Archaean in age[2]. Later investigations have proved that much of this gneiss, is not of Archaean age, but is of intrusive origin and has invaded rocks of various ages at a number of different geological periods.

Three kinds of granite have been recognized in the Archaean complex of the Himalayas. These granites are: biotite-granite, hornblende-granite, and tourmalin-granite. Of these the most prevalent is the biotite-gneiss or granite. The composition is acidic and the bulk of the gneiss is made up of milk-white orthoclase. Next to the gneisses, the most frequent rock is biotite-schist. These rocks are abundantly traversed by dykes, stocks and masses of basic intrusives such as dolerite, epidiorite, gabbro, etc.

The spatial distribution of these gneissic rocks lies mainly in the Zanskar Range and the region beyond it, in Gilgit, Baltistan and Ladakh, while in the ranges to the south of the Valley of Kashmir they are not so pronounced. The core of the Dhauladhar Range is formed by these rocks, but they are not very conspicuous component of the Pir Panjal Range, where they occur in a number of minor intrusions. The trans-Jhelum continuation of this range, known as Kaz-nag, has a larger development of the crystalline core. A broad area of Kishtwar is also occupied by these rocks[3].

Dharwar System

Dharwar System is used as synonymous with metamorphosed Archean sediments. These sedimentary strata appear to rest over the gneisses at some places with an unconformity, while at others they are largely interbedded with them. Dharwar strata are altogether unfossiliferous. The Dharwar rocks are highly metaliferous, containing ores of iron-ore, and manganese and occasionally also of copper, chromium, tungsten, lead, zinc, silver and gold. These rocks are also rich in mica, corundum, lithium, titanium, thorium, lepidolite, ruby, sapphire, emerald, zircon, garnets, granites, marbles, ornamental building stones, and roofing slates. The famous marbles of which the best specimens of ancient and medieval Indian architecture built are the products of the Dharwar System.

The Dharwarian rocks in the state of Jammu and Kashmir are found mainly in Ladakh, Gilgit, Kohistan and North Hazara. They consist of slates, phyllites (often graphitic), schists, migmalite, quartzites and crystalline limestones. They have been named as *salkhala-series* in the Kashmir Division, *Jutog-series* in the Shimla area, and *Daling series* in the eastern Himalayas. The great Himalayan range, west of Ladakh, is largely composed of the Salkhalas converted into para-gneiss, the Nanga Parbat (8,119m) massif

FIGURE 1.2

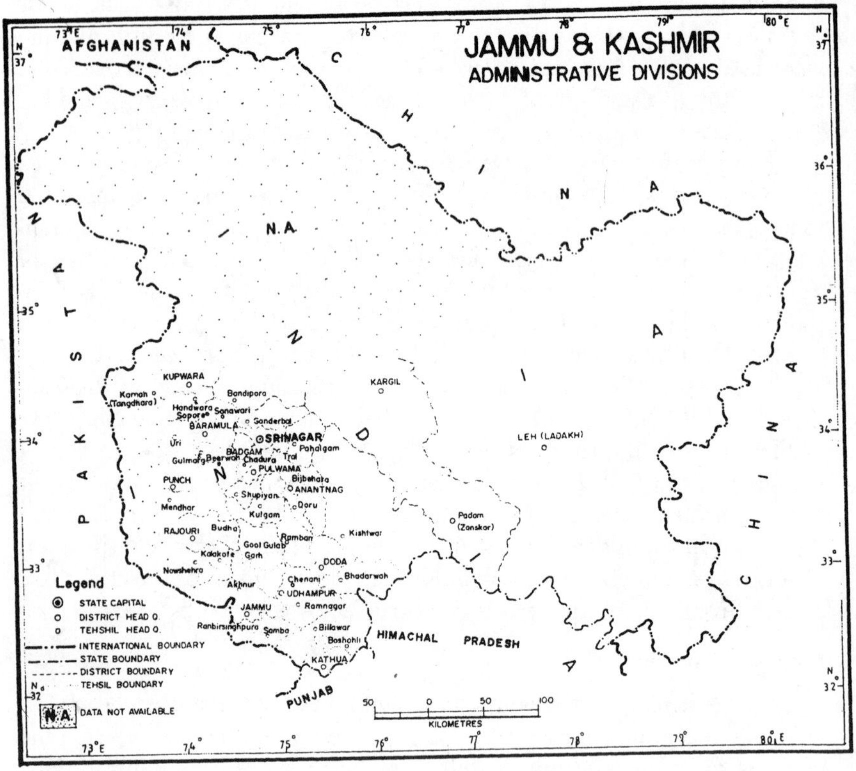

being almost wholly built of this with intrusive biotite-gneiss and hornblende-granite rocks (Fig. 1.2). The Salkhala sediments have been subjected to an intense granitisation at places, in Kaghan, in the ranges north of the Kishan Ganga river, and in the Nanga-Parbat area[4].

The Cambrian System

Cambrian (from the Roman name for Wales-Cambria) is the oldest system of rocks in which fossils can be used for dating and correlation. The period commenced at lest 530 ± 40 million years ago, and had a duration of at least 70 million years. The Cambrian rocks cover an extensive tract in Handwara, in the Baramulla district to the north of the Jhelum. They are also found in Pir Panjal, the Kishenganga Valley and in Hazara (Fig. 1.2).

The Mesozoic (middle life) Era ranging from 230 to 70 million years, comprise the Triassic, Jurassic, and Cretaceous Systems. The Mesozoic war preceded by the palaeozoic and followed by the Tertiary.

Of the Mesozoic System, the Triassic is the best and most fully developed in the state; Jurassic and Cretaceous outcrops are few and mostly confined to the mountains of Ladakh which have not been fully surveyed by the geologists. The most spectacular elements of Mesozoic fauna were the giant reptiles, the dinosaurs, which reached their maximum size in the Jurassic and occupied most of the ecological niches. The flora of the Jurassic included many forms still living at the present day such as conifers and ferns. The Jurassic formation is found in the Banihal Valley. In the Liddar Valley, Middlemiss has proved a continuous succession of fossiliferous palaeozoic strata of the Cambrian, Ordovicaian and Silurian, Devonian, Carboniferous and Peremian[5]. The Ordovicaian rocks are found near Traehgam (Hundawara Tehsil) of the Kupwara District of Kashmir.

Round the oval expanse of the core of the Liddar anticline, there runs a thin but continuous band of Silurian strata. The aggregate thickness of the fossil-bearing Silurian strate is only 30 m, but the organism preserved in them leave no doubt of their age, thus denoting a highly valued geological horizon in India. The occurrence of Silurian rocks is suspected, on strong lithological grounds, in Poonch and in Chitral[6].

The Devonian rocks are found on both the flanks of the Liddar anticlinal as thin bands. The Devonian rocks have great thickness of massive white quartzite.

The strata of Carboniferous (Syribgothyris Limestone Series) is found around Aishmuqam and Kotsu (Kashmir Valley) which have numerous fossils. The limestone of the Syringothyris series of considerable thickness, (600-900 m) are observed in the Banihal Valley of the Pir Panjal.

Gondwana of the Himalayas

At several localities along the foot of the Himalayas (Hazarato Assam) strips of Lower Gondwana rocks are found. Only in the Kashmir areas these rocks attain any development and exhibit normal stratigraphic relation to the parmo-carboniferous. The rich strata of the Middle Gondwana with their coal seams are, however, absent in the state of Jammu and Kashmir and the North-western Himalayas.

The Upper Carboniferous and Permian Systems

The Upper Carboniferous and Permian Systems are well developed from Kashmir to Kumaon. They are well developed in Kashmir Division also.

The Upper Carboniferous consists of a thick (over 2,400m) volcanic series—*Panjal Volcanic Series*—of bedded tuffs, slates, ash-beds and andesitic to balastic lava flows (*Panjal Trap*). The slaty tuffs contain at places marine fossils allied to the fauna of the productus limestone.

The mid-Palaeozoic land mass of Kashmir performed one important function: it must have served as a land-bridge between Gondwana and the great northern Eurasian continent (Angaraland). It was through this land-bridge that the terrestrial vegetation of the Indian portion of Gondwanaland established some links with angaraland.

In the Upper Carboniferous System and metamorphic belt of the N.W. Himalayas, extending from Kaghan to Jammu, a voluminous series (upto 1000 m. thick) of metamorphosed rocks of markedly arenaceous composition-banded argillaceous, quartzites, grits, phyllites and quartz-schist, occurs in a number of fold-faulted, disturbed longitudinal basins, one four miles across the strike. These have been named from the Tanawal country in Hazara. In the Poonch Pir Panjal these rocks show a clear lateral passage into the *Agglomeratic Slato Series* of Upper Carboniferous age. The whole group is entirely devoid of fossils[7].

The Triassic System

The Triassie period extends from 225 to 195 million years, a duration of 30 million years. It marks the beginning of the Mesözoic Era. The *Trias* is one of the best developed sedimentary formations of the Inner-Himalayas from Hazara to Nepal. In Kashmir, a thick series of compact blue-limestone, slates and dolomites is conspicuously displayed in many of the hills bordering the valley to the north, while they have entered largely into the structure of the higher parts of the Sind, Liddar, Gurez and Tilel valleys of the north-east flanks of the Pir Panjal. A superb development of limestone and dolomites of this system is exhibited in a series of picturesque escarpments and cliffs forming the best part of the scenery north of the Jhelum. Another locality for the development of the *Trias*, is the Pir Panjal, extending from beyond the Jhelum to Kishtwar. A great part of the Triassic on the north-east flanks, however, is obscured under later formations such as the Karewas and moraine debris[8].

Limestones are the principal components of this system. The rocks are of light blue or grey tint, compact and homogenous, and sometimes dolomitic in composition. They compose a very picturesque feature of the landscape. Numerous springs of fresh water issue from the cliffs and prominence of theirs limestones at the south-east end of the valley, and form the sources of the Jhelum, the best known of these are the river like fountains of Achabal and Verinag and the springs of Anantnag and Bhawan.

The Middle Trias is visible at many points at Vihi, Khrew, Pastanah, Khunmu above Pahalgam in the Liddar and in some of the tributary valleys of the Upper Sind.

The Upper Triassic limestone is widely distributed in moraine heaps clothing the north-eastern slopes of the Pir Panjal. It is also from north Pahalgam, through the head-waters of the Sind, to beyond Gurez[9].

The Triassic limestone has furnished an abundant building material to the architects of ancient Kashmir in the buildings of their great temples and edifices, including the famous shrine of Martand.

The Jurassic System

Named from the Jura Mountains of France (by von Humboldt in 1795), the Jurassic System is the period of time extending from 195 million years, having a duration of about 60 million years. The fauna of the Jurassic was extremely diverse and varied. The dominant terrestrial animals were the dinosaurs, which reached their maximum size in the Jurassic and occupied most of the ecological niches. The first bird appeared in the Upper Jurassic, but the mammals were an insignificant element of the fauna and were rarely larger than the modern rat. The flora of the Jurassic included many forms still living at the present day, such as cycads, ginkoes, conifers, and ferns.

In the Zanskar range of Spiti, the Upper Trias is succeeded by a series of limestones and dolomites of a great thickness (Fig.1.1). This is known as Kioto limestone which is overlain by the most characteristic Jurassic formation of the Inner Himalayas, known as the *spiti shales*. An outcrop of the Jurassic System is found on the north side of the Banihal-pass of the Pir Panjal. It is probable that similar outliers of the Jurassic exist in association with the extensive Triassic formations of the northern flank of the Pir Panjal between Banihal and Gulmarg, under cover of the Pleistocene glacial and karewa deposits. These rocks are well displayed on the north side of the Sind above Sonmarg and in the Amarnath-valley[10].

Cretaceous System

Cretaceous Period (Greek *creta, 'chalk'*) has a duration of approximately 72 million years. The division of the period into Lower and Upper is at about 100 million years. The Lower Cretaceous sediments continue the pattern of Jurassic sedimentation. The Upper Cretaceous is represented by the unique white limestone as the chalk. The end of the Cretaceous times marks the end of the Mesozoic Era[11].

Rocks of the Cretaceous System are found in a few localities in Kashmir. They are mainly confined either on the Great Himalayan Range between the Burzail and Deosai or in the Zanskar Range in the Rupshu province[12].

Two or three small patches of Cretaceous rocks occur in Rupshu which correspond to the Chikkim-series in their geological relations. They are composed of a white limestone, forming some of the highest peaks of the Ladakh-range. A highly interesting group of volcanic rocks—laminated ash-beds, tuffs, agglomerates, coarse agglomeratic conglomerate and bedded balastic lava-flows, associated with marine Cretaceous limestones on the one hand and with a varied group of acid and basic plutonic intrusives—granites, porphyries, gabbro and peridotite—on the other[13]. These rocks are folded into a synclinal trough lying among Salkhalas, extending from south-east of Astor to beyond Dras, traversing the Great Himalayan Range at the head of the Burzail Valley in a 20 km wide outcrop. The stratified volcanic series, several thousand feet in thickness, contains numerous sedimentary layers, limestones and shales. South of Ladakh, running along the Indus Valley is a band of Cretaceous rocks. Middle and Upper Cretaceous sediments are found in Chitral.

The Tertiary System

The period of time which elapsed between the end of the Cretaceous and the present time, having a duration of 65 million years, to 10 million years is the Tertiary System. The division of the Tertiary System may be made as follows:

Holocone (Youngest)
Pleistocene
Pliocene
Miocene
Oligocene
Eocene
Palaeocene (Oldest)

The Tertiary Era is the most important in the physical history of the whole Indian subcontinent. It was during these ages that the most important surface features of the area were acquired, and the present configuration of the country was utlined. The pile of marine sediments, that was accumulating on the border of the Himalayas and in the Tibet since the Permian Period, began to be upheaved by a slow secular rise of the ocean bottom. There appear to have been three important phases of upheaval of this mountain system[14].

The Tertiary band at the Jhelum stretches eastwards through the Kashmir area, preserving all its geological characters and relations unchanged to the Ravi and thence to the Sutlej[15].

The Tertiary outcrop is the widest where it crosses the Jhelum. The Tertiaries of the Inner Himalayas (*The Indus Valley Tertiaries*) are found in south Ladakh. The Ladakh Tertiaries extend from Dras, Kargil, and Leh to beyond Hanle, north of Spiti.

The Tertiary zone of the Outer Himalayas is disposed in three or four parallel belts conforming to the general strike, the oldest, Eocene, abutting on the south-western flank of the Pir Panjal, while the newer ones build the successive outer low ranges of the foothills zone.

The Eocone rocks which stretch from Kohat through the Hazara mountains to Kashmir, persist as a narrower band across the Jhelum and runs along the foot of the Pir Panjal for more than 400 km (Fig.1.1).

The Subathu series of Jammu are exposed near Riasi, north of Jammu. Another is seen in Punch exhibition like relations. At Riasi the overlying coal-measures containing seams of anthracitic coal upto 6 m in thickness, have been found to be workable and capable of supporting remunerative mining, but the westward coal-seams become thinner and then disappear[16]. The Oligocene is not fully developed in India, but the Miocene rocks have great stretch in the Sind, Murree, Potwar and Kashmir[17].

The Siwalik System

The new Tertiaries occur on an enormous scale, forming the low outermost hills of the Himalaya along its whole length from the Indus to the Brahmaputra. They are known as the Siwalik System, because of their constituting the hills near Hardwar where they were first known to the scientists, and from which were obtained the treasures of fossils[18].

The Siwalik System is a great thickness of detrital rocks, such as coarsely-bedded sandstones, sand-rock, clays and conglomerates, measuring between 4,500 to 5200 m in thickness. The inner boundary of the Siwaliks is faulted one only as far as Chenab river, beyond which westwards, the fault gradually diminishes and is replaced by an anticlinal flexure. It is well-marked and typical at Udhampur, but has lost is significance at Kotli where Siwalik outliers are found inside the boundary of the Siwalik zone in synclinal troughs of Murees[19]. The weathering of the Siwalik rocks has been proceeding at an extraordinarily rapid pace. There are gigantic escarpments and dip-slopes, separated by broad longitudinal strike valleys and intersected by deep meandering ravines of the streams. The strike is markedly constant in a N.W.-S.E. direction (Fig.1.3)

The Siwaliks with a width of about 40 km are in parallel folded zones in the Jammu Division. The Upper and the Middle Siwaliks of the Jammu hills show similarity in lithology with the Salt-Range and Potwar areas[20].

The Pleistocene System

The Pleistocene period of geology is the most fascinating, though the briefest of earth history. It was during thin period that the geography and topography of most parts of the world acquired their final outlines. The pre-

FIGURE 1.3

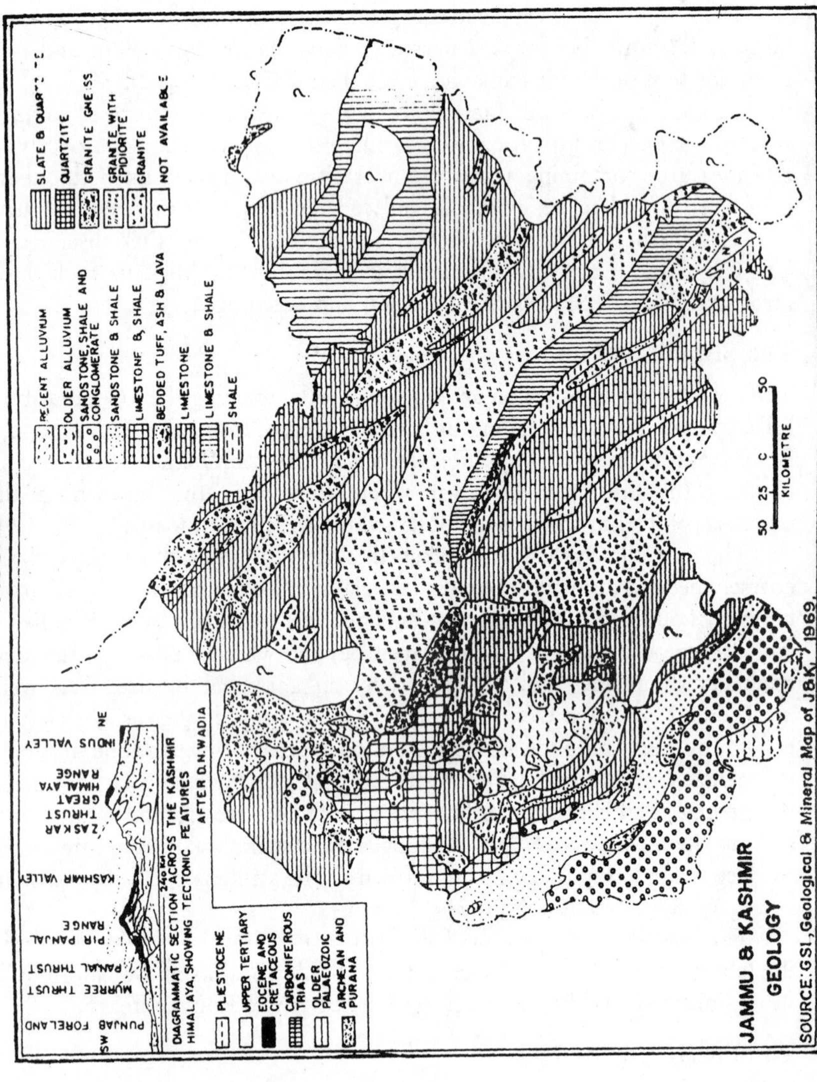
JAMMU & KASHMIR
GEOLOGY
SOURCE: GSI, Geological & Mineral Map of J&K, 1969.
RECENT ALLUVIUM
OLDER ALLUVIUM
SANDSTONE, SHALE AND CONGLOMERATE
SANDSTONE & SHALE
LIMESTONE & SHALE
BEDDED TUFF, ASH & LAVA
LIMESTONE
LIMESTONE & SHALE
SHALE
SLATE & QUARTZITE
QUARTZITE
GRANITE GNEISS
GRANITE WITH EPIDIORITE
GRANITE
NOT AVAILABLE
50 25 0 50
KILOMETRE
SW
NE
PUNJAB FORELAND
MURREE THRUST
PANJAL THRUST
PIR PANJAL RANGE
KASHMIR VALLEY
ZASKAR THRUST
GREAT HIMALAYA RANGE
INDUS VALLEY
240 KM
DIAGRAMMATIC SECTION ACROSS THE KASHMIR HIMALAYA, SHOWING TECTONIC FEATURES
AFTER D.N. WADIA
PLIESTOCENE
UPPER TERTIARY
EOCENE AND CRETACEOUS
CARBONIFEROUS TRIAS
OLDER PALAEOZOIC
ARCHEAN AND PURANA

sent distribution of fauna and flora also occurred during the Pleistocene Period. The Pleistocene system in India has a fuller and more varied development than all the preceding system save the Archaen; it covers about 650,000 sq km of North India.

The Pleistocene deposits are widespread in many parts of Kashmir. These deposits are known as Karewas (Wudur). The Karewa formation occupies nearly half the area of the valley. The Karewas have a width of from 13 to 26 km along its south-west side and extends for a length of some 80 km from Shopian to Baramulla. The present view regards the Karewas as the surviving remnants of deposits of a lake or series of lakes which once filled the whole valley-basin from end to end[21]. The highest limit at which the karewas have been observed on the north-eastern slopes of the Pir Panjal is 3,800 m, more than 2,000 m above the level of the Jhelum bed. At the height of the Ice-Age this Karawa Lake must have been no less then 7,800 sq km in area[22].

The Karewas are mostly horizontally stratified deposits formed of beds of fine-grained sand, loam and blue sandy-clay with lenticular bands of gravelly conglomerate, fossils of birch, beech, willow, oak, walnut, trapa, rose, pines and vertebrates, and fish are found in the karewas.

Pleistocene and later glacial deposits are widely distributed in Kashmir which may be observed in the Sind, Liddar, and Lolab valleys. Moraines, belonging to three or four successive glacial advances, interbedded with the karewa deposits at various levels have been recognized by de Terra[23].

Among the later deposits then these are the high-level river-terraces, 300m or more above the stream-bed which include the enormous fans in the Nubra, Changchenmo valleys of Ladakh and the cave deposits of Harwan (Srinagar)[24].

The karewas, in their upper part at least, are supposed to be the relics of old extensive lake basins, which intermittently came into existence during the warm interglacial periods of melting ice and which periodically filled the whole Valley of Kashmir from end to end to a depth of more than 300 m. This old alluvium has been subsequently elevated, dissected, and in a great measure removed by subaerial denudation as well as by the modern Jhelum, leaving the karewa outliers of to-day[25]. A synoptic view of the Kashmir Valley during the Pleistocene Period has been given in Fig. 1.4.

Physical Features

Physical features of a region largely influence the climate, soils, vegetation and lifestyle the people. An understanding to physical features is imperative to assess the resource potential and their management. In the following paragraphs a brief account of the relief or contours, mountain systems, glaciers,

FIGURE 1.4

KASHMIR VALLEY
DURING LATE MIOCENE
(7 Million year)

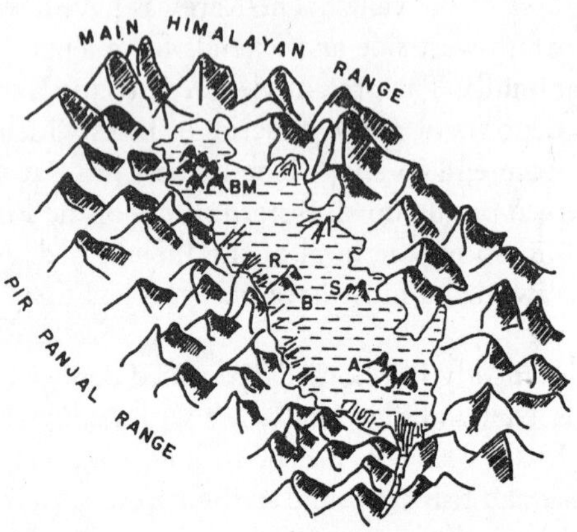

A............ ANANTNAG
B............ BEERWA
BM......... BARAMULLA
R........... RAFIUDDIN HILL
S........... SOBUR

ORIGIN : REMNANT OF PLETISTOCENE LAKE OR OX-BOW LAKE

earthquakes, drainage system, and springs has been given.

On the basis of geomorphic features the state of Jammu and Kashmir is divisible into three distinct physical regions (Fig.1.5).

1. The Outer Plains
2. The Siwaliks
3. The Lower or Middle Himalayas
4. The Valley of Kashmir
5. The Greater Himalayas
6. The Plateau of Ladakh

1. The Outer Plains (Andarwah and Bajwat)

The outer plain of Jammu and Kashmir, locally known as Andarwah or Bajwat, is a part of the Great Plains of India. In the opinion of Oldham, the depth of the alluvium of the Indo-Gangetic plain is about 4,600 m near its

FIGURE 1.5

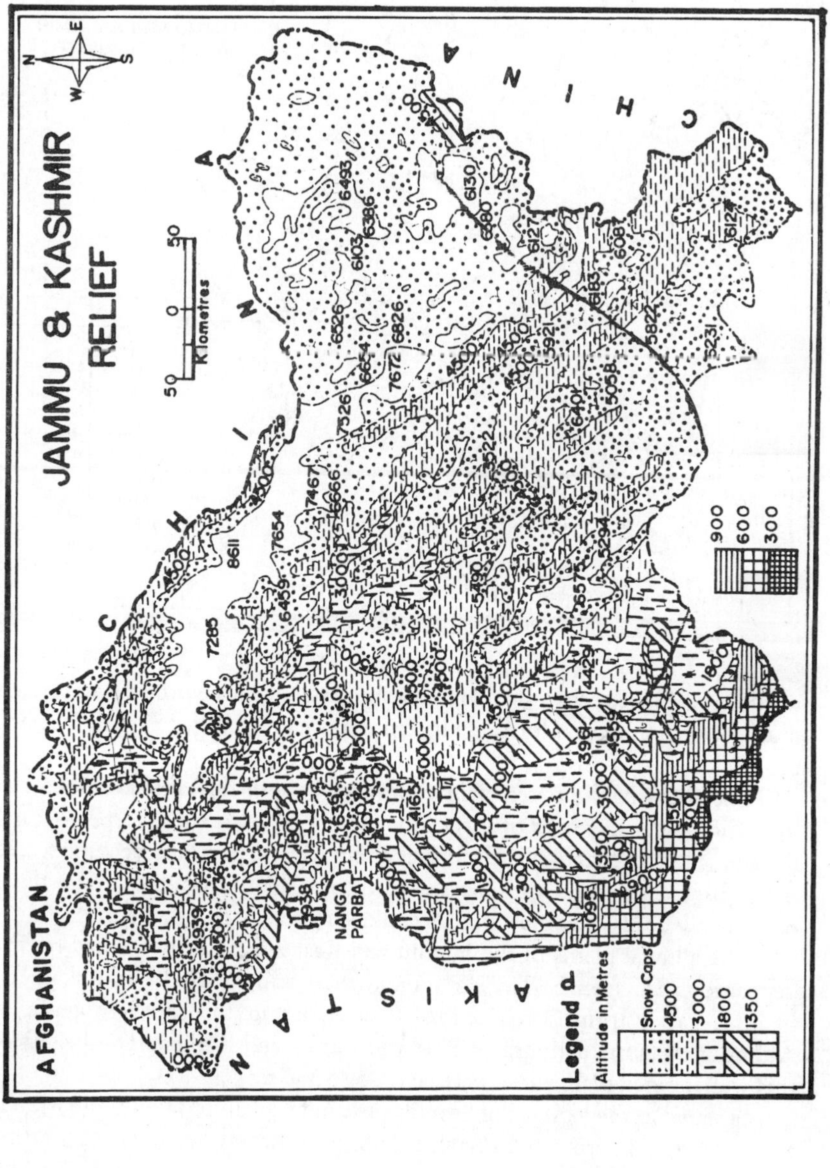
JAMMU & KASHMIR
RELIEF
50 0 50
Kilometres
AFGHANISTAN
PAKISTAN
CHINA
CHINA
NANGA PARBAT
Legend
Altitude in Metres
Snow Caps
4500
3000
1800
1350
900
600
300

FIGURE 1.6 (a)

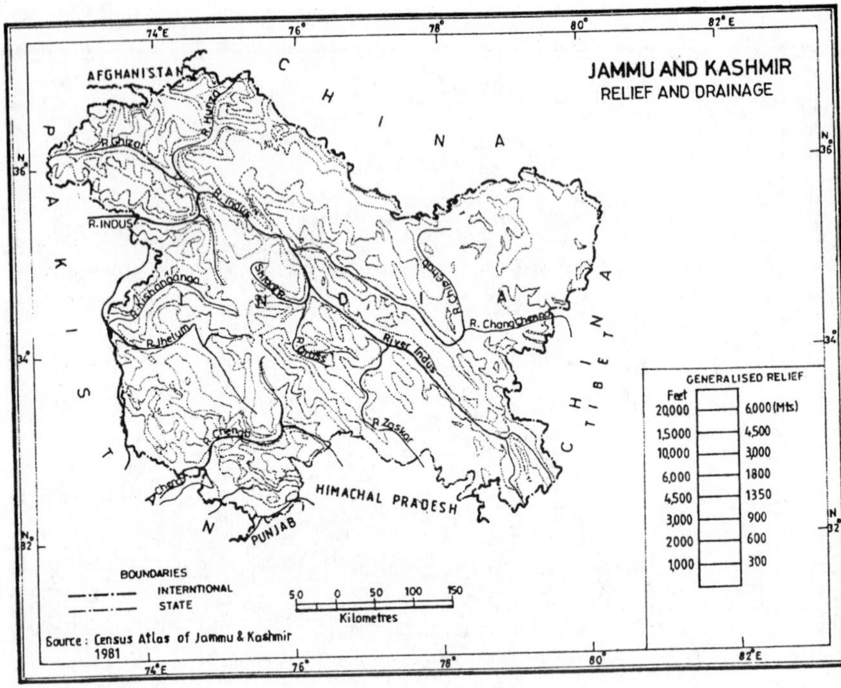

northern limit. Recent calculations from geodetic survey, however, give a much greater thickness for these lighter deposits resting on the pre-Tertiary bed-rock[26].

The rocks of the outer plain are everywhere of fluviatile and subaerial formation—massive beds of clay, either sandy or calcareous, corresponding the silts, mud and sand of the modern rivers. Gravel and sand become scarce as the distance from the hills increases. At some depth from the surface there occur a few beads of compact sands and even gravelly conglomerates.

The alluvial plains of the Jammu and Kashmir state are formed by the depositions of Chenab, Tawi, Ravi and their tributaries. In width, in the plain it varies from 7 to 30 km. It stretches in the southern parts of Akhnoor, Ranbirsinghpura (R.S. Pura), Samba and Kathua. The height of this well-drained plain varies between 325 to 350 m[27].

Situated in the vicinity of Siwaliks, the outer plain of Jammu Division is traversed by numerous hill torrents, locally known as *Khad.* These torrents remain dry over the greater parts of the winter and summer seasons and attain enormous size during rains. At the occurrence of rains, the ravines

FIGURE 1.6 (b)

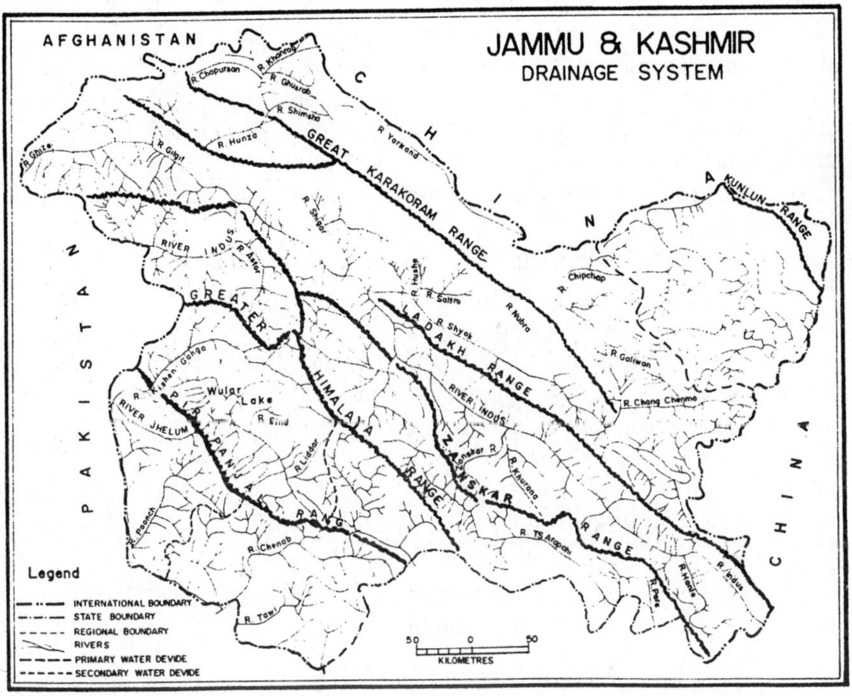

achieve the shape of gushing torrents and it becomes difficult to ford them. Occasionally, at the occurrence of floods these torrents transport massive quantities of sand, red silt, pebbles and shingles. The channel of these torrents are characterized by clumps of reed growth known as *Khar*.

2. *The Siwaliks*

The relatively low outermost hills of the Himalayas along its whole length from the Indus to the Brahmaputra are known as the Siwaliks. The Siwaliks are conspicuously well developed for a distance of about 200 km between the Ravi and the Jhelum rivers in the state of Jammu and Kashmir. The width of the Siwaliks is between 20 to 50 km and height above the sea level between 600 to 1220 m in the state of Jammu and Kashmir. The rocks of the Siwaliks are disposed in parallel folded zones. The outer most ranges of the Himalayas rise from the plains of Punjab, commencing with a gentle slope from Jammu, attain about 600 m in altitude, and then end abruptly in steep, almost perpendicular escarpments inwards. Then follows a succession of narrow parallel ridges hill, their strike persistent in a N.W.-S.E. direction, separated by more or less broad longitudinal or strike valleys (the basin of subsequent streams). These valleys are known as *duns.* In the Jammu Hills the extensive, picturesque *duns* of Udhampur, and Kotli are quite typical. The Siwaliks are made of the younger Tertiary rocks[28].

The southern slopes of the Siwaliks are gentle, covered with deciduous forests, evergreen bushes and numerous scrubs. The northern slopes are steep with thin cover of vegetation. In the Siwaliks to the east of Jammu City are situated the Mansar and Saruinsar lakes at an altitude of about 600 m above the sea level.

The undulating slope, adjacent to the plains upto an elevation of 300 m between Ravi and Chenab rivers are locally known as *Kandi.* The *Kandi* land is characterised by numerous torrents, hilly soils and scarcity of water. Consequently, agriculture in Kandi tract is largely dependent on rains. The rate of soil erosion during rainy season is heavy. Cultivation of maize, millets, pulses, oilseeds, and fodder, and rearing and keeping of cattle are the dominant primary economic activities in Kandi region. The Gujjars are the buffalo-keepers who are called as *Doodhia-Gujars.* Economically, this is the most underdeveloped part of the state in which a substantial proportion of population (40%) is living below the poverty line.

The second largest city and winter capital of the state, i.e., Jammu is situated at an altitude of 366 m on the southern slope of the Siwaliks.

3. *The Middle Himalayas (Lesser Himalayas)*

The Middle Himalayas of the Jammu and Kashmir state lie between the Ravi in the east and the Punch in the west and continues upto Muzaffarabad. The ranges of Pir Panjal and Dhaula dhar lie in the Lesser Himalayas of the state. In Jammu region they are locally known as *Pahar* (mountain). In elevation they seldom rise above 3600-4600 m, and have a width of about 60 km near Rajauri. In general, it has an east-west extension. At Tanger, as a branch ridge runs due south-west. This branch in known as Trikuta Mountain, which is characterised with limestone strata. The sacred cave of Vaishno Devi is situated on a peak of Trikuta Mountain. The Lesser Himalayas form an intricate system of ranges. The Pir Panjal is a singularly well-defined range form Kaghan Valley to beyond the Ravi Valley, which may be taken as a type of mountain of the Middle Himalayas, these rides present generally a steep escarpment towards the plain and a long gentle slop towards Kashmir. Such mountains are spoken of as having an 'orthoclinal' structure, with a writing desk. To this course (among several others) is due to the presence of dense forest vegetation, the glory of the Middle Himalayas, clothing the northern and north-eastern slopes, succeeded higher up by a capping of snows, while the opposite, southern slopes are, except in protected valley-slopes, barren and devoid of snow, being too steep to maintain a soil-cap for the growth of forests or allow the winter-snows to accumulate[29].

The Middle Himalayas are composed of highly compressed and altered rocks of various geological ages, ranging from the Puranas and Carboniferous to Eocene[30].

Several important rivers, like Tawi, Manawar-Tawi, Basantra, and Ujh have their sources in the Middle Himalayas. The rock strata of the Middle Himalayas is dominated by sandstone, shale, and light yellow limestone of the Miocene Murree System and Eocene deposits.

Excepting the river terraces and the river valley, cultivation of crops is not an important economic activity on the slopes of the Middle Himalayas. The people are largely dependent an forestry, lumbering, herding, cattle keeping, and tourist industry. The general standard of living of the people is low and the rural areas are less accessible. The area is poor in basic mineral resources, specially in coal, petroleum and natural gas. Coal, however, being mined at Kalakote (Dist. Udhampur) of the Jammu Division. The Middle Himalayas are very rich in clays, sands, limestone, building-stone, slate and have enormous water resource which is being harnessed for the generation of hydel power. There is however, a dearth of metallic minerals like iron-ore, copper, bauxite, pyrite, lead, zinc, silver and gold, in the Middle Himalayas.

4. *The Valley of Kashmir*

The Valley of Kashmir, nestled in north-western folds of the Himalayas is a transverse valley, surrounded on all sides by high mountain ranges, characterized by snow covered lofty peaks. The surrounding mountain ranges, rising to a height of 5,550 m on the north-east, where the Banihal-Pass (Jawahar-Tunnel) provides an exit from the valley. The only outlet for the rivers is the Baramulla-gorge, where the placid Jhelum river leaves the smooth grassy banks and hurries headlong down its rock course to the great plains of the south.

In the opinion of Lydekker, the Kashmir Valley is distinctly basin-shaped, and that it has a length of about 140 km (84 miles) and a width varying from 53-55 km (20 to 25 miles). The lowest point in the valley has an elevation of 1600 m (5,200 feet) and the mean elevation is 184 m (6,000 feet) above the sea level[31].

In its course the river Jhelum, below the town of Anantnag (Islamabad), flows through a plain of low level recent alluvium: the width of this plain varies from two to fifteen miles. It appears level to the eye having in the first 60 km (36 miles) a fall of 51 m (165 feet) but only 17 (55 feet) in the lower four and a half miles. This alluvium has been formed by the river in flood,

FIGURE 1.7

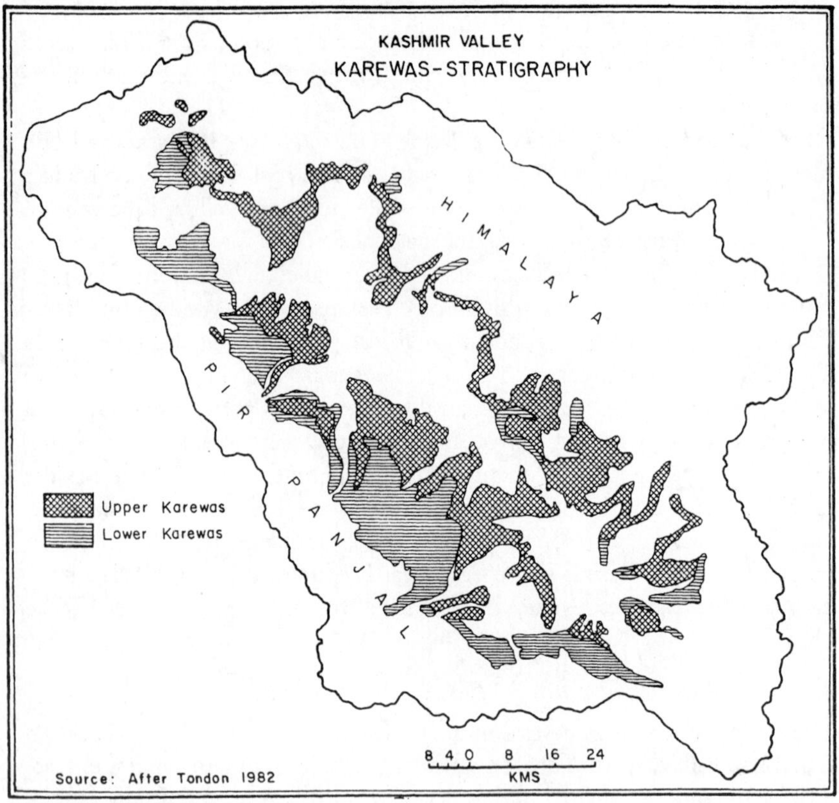

and its formation may still be locally observed, though the operations of natural agencies have been greatly impeded by artificial embankments[32].

It is chiefly composed of loam and clay, and it would be difficult to distinguish it from the deposits now forming in the lakes of the valley, though the latter may be more distinctly stratified. There is no evidence to show that any change of level has taken place since the deposition of this alluvium, which contains in many cases pottery and other works of art, of a comparatively modern date[33].

The oval shaped Valley of Kashmir is filled with thick deposits of alluvium which have blanketed even the lower slopes of the surrounding ranges. On the borders of the Kashmir Valley and even in the valley floor occur extensive elevated plateaus of alluvial and lacustrine material. These deposits are locally known as *karewas* (Fig.1.7). Owing to their elevated position these karewas cannot generally be brought under irrigation. They are, therefore, in summer, easily distinguished from the plain of the river alluvium, which in most part is densely covered with rice crop. In the central parts of the valley the karewas consist chiefly of loam, or loamy clays with, but faint indications of stratification and with level surfaces[34].

The karewas are divided from each other, sometime cut into stripes by ravines of from 30 to 92 m (100 to 300 feet), in depth; occasionally, they surrounded altogether by lower ground, but more generally they connect on to some of the mountains that bound the valley. Karewas occupy a width varying from 12 to 25 km (8 to 16 miles) along the south-western side of the valley, for a length of about 83 km (50 miles) from near Shopian to the river flat between Sopore and Baramulla. On the north-east side of the valley across the river, on it right bank, are small tracts of karewas. The karewas adjoining the mountains have their surfaces inclined from the latter with decreasing slope. On the south-eastern side of the valley the karewas reach upwards to an elevation of about 2,000 m (6,500 feet) or 400 m (1,300 feet) above the lowest part of the plain of the river alluvium (Fig.1.7)

The flat-topped karewas always consist of horizontal beds, and in the neighbourhood of Anantnag (Islamabad) attain a thickness of about 92 m (300 feet).

The sloping karewas are found along the flanks of Pir Panjal Range, where they form a continuous series, reaching from Shalura in the north-west to below Shopian in the south-east (Fig.1.7).

At the mouth of the Sind Valley, on the north-eastern side of the Kashmir Valley, there are thick lacustrine deposits, consisting of conglomerates, sand and gravel. They have a similarity with the deposits of Anantnag (Islamabad) and the Pir Panjal.

Side Valleys

Within the Valley of Kashmir, there are three important side valleys, namely: (i) The Liddar Valley (ii) The Sind Valley, and (iii) The Lolab Valley. A brief description of these valleys has been given below:

Liddar Valley

The Liddar Valley extends from Anantnag (Islamabad) to Pahalgam, including the Parganas of Dachinpora and Khaurpora. At Pahalgam, the valley divides itself into two branches which stretch obliquely, one towards the north-east, leading up to the Sheshnag and the Cave of Amarnath. At its lower end the valley is about 5 to 8 km wide but at its upper end it is only a few hundred metres in width. On both sides, it is bounded by lofty mountains covered with dense forests. Cultivation of crops does not extend beyond about a mile north of Pahalgam. There is a road along the left bank of the river upto the Amarnath Caves. This path is yearly traversed by lakhs of Hindu pilgrims, in the months of July and August.

The river Liddar is fed with streams from the Kolahai and Sheshnag. Towards Kolahai is a small valley covered with grasses where glades are used by graziers. The steep sides of this valley show marred ice action with several waterfalls till the Kolahai glacier at an altitude of about 5470 m (17,779 feet).

Sind Valley

It is one of the most beautiful valleys of Kashmir. It owes its name to the river Sind-Lar which is fed by glaciers and streams. It extends from Ganderbal to the Zoji-La Pass, covering a distance of about 100 km. Sonamarg (golden meadow) is a beautiful meadow near the Thajwas Glacier.

The Sind valley widens after Shadipur. On either side of the valley are the lovely snow covered peaks. The valley abounds in large forests of pine, and other fruit trees. The famous lakes of Vishansar, Krishnsar and Gangabal are within the catchment of Sind river which attract large number of tourists every year.

Lolab Valley

This fertile oval shaped valley is situated in the north-west side of Kashmir, it is about 25 km long and 5 km wide. It is traversed by the Lahwal Stream which flows down from the surrounding hills. The valley is covered by thick forest of deodar. The extensive pasture lands in the valley are the excellent grazing grounds. It is also famous for apple, cherry, peach and walnut.

5. The Greater Himalayas or Inner Himalayas

To the north of Pir Panjal and Dhauladhar ranges are the more lofty mountain ranges of the innermost zone of the Himalayas rising above the snow-line into peaks of perpetual snow. In the north, Kashmir Range, is an offshoot of the Zanskar Range, which forms the north-eastern border of the Valley of Kashmir. In the Zanskar Range there are a few peaks which have heights between 4,500 to 6,100 m above the sea level. Beyond this range, the country, with the exception of the deep gorges of the Middle Indus, is a high-level plateau-desert, utterly devoid of all kinds of vegetation. Here, one may observe the elevated plateaus and high mountain ranges separated from one another by great depressions, with majestic peaks towering to 7,300 m. The altitude steadily increases farther north, till the Peak K^2 on the mighty Karakoram or Mustang Range, attains the culminating height of 8,621 m, the second highest peak in the world. Farther north is the Karakoram Range which makes a watershed between India and Tajkistan. The valleys of these regions show varying characters. In the south-east is the Changchenmo whose width is from eight to ten km with an average height of 4,270 m above the sea level. From that to the north-west the height of the valley beds descends, till in Gilgit on the very flanks of the gigantic peak of Nanga Parbat, Diyamir (8,120 m), the rivers have cut so deeply through the bare, bleak mountains that the streams flow at an elevation of only 1,525 m above the sea level. At places in north and north-east Kashmir, there are extensive flat, wide plains or depressed tracts among the mountains, too wide to be called valleys, of which the most conspicuous are the plateaus of Deosai, 3965 m. high, and Lingzhitang, 4,880 m. The physical features of this extremely rugged, wind-swept and frost-bitten region vary much in character. They present an aspect of desolate, ice-bound altitudes and long dreary wastes of valleys and depressed lands totally different from the soft harmony of the Kashmir mountains, green with the abundance of forest, pastures and cultivation[35].

6. Ladakh Plateau

Ladakh is one of the loftiest inhabited regions of the world, (3,600-4,600m). Its short but warm summers enable a few grain and fruit crops to ripen. Owing to the great aridity of the atmosphere, the climate is one of fierce extremes, from burning heat of some of the deserts to several degrees below freezing-point at night. The bare mountains which rise from them exhibit the exquisite from burning heat of some of the deserts to several degrees below freezing-point at night. The bare mountains which rise from them exhibit the exquisite desert coloration of the rocks due to the peculiar solar weathering. Between Ladakh and the Dhauladher-range are the districts of Zonskar, Lahoul, and Rupshu, consisting of intricately ramifying glaciated ranges of crystalline rocks, intersected by lofty valleys having but a restricted drainage into a few saline lakes and marshes[36].

As stated in the preceding paragraphs, the average height of the Greater Himalayas extend to 6,100 m. On it are situated the peaks, like Mount K^2—8615 m (Karakoram), Nanga Parbat—8,119 m (Kashmir Himalaya), Gasherbrum—8,073 m (Karakoram).

Geological Classification of the Himalayan Ranges

As regards geological structure and age, the Himalayas fall into three broad stratigraphical belts or zones. These zones do not correspond to the geographical zones, given above, as a rule[37].

1. The Northern or Tibetan Zone

The Tibetan zone lying behind the line of highest elevation, (i.e., the central axis corresponding to the Great Himalayas). This zone is composed of a continuous series of highly fossiliferous marine sedimentary rocks, ranging in age from the earliest Palaeozoic to the Eocene age.

2. The Central or Himalayan Zone

The Central or Himalayan zone comprising most of the Lesser or Middle Himalayas together with the Greater Himalayas. It is mostly composed of crystalline and metamorphic rocks-gneisses and schists, with unfossiliferous sedimentary *Purana* and Mesozoic deposits.

3. The Outer or Sub-Himalayan Zone

The Outer or Sub-Himalayan zone corresponds to the Siwalik Ranges, and composed almost entirely of Tertiary, and principally of Upper Tertiary, sedimentary deposits.

Significance of the Himalayas

The mighty range of the Himalayas exercise great influence over the climatic condition, economy and society of the Indian subcontinent. It closely influences both its air and water circulation. Its mighty snowy ranges have a moderating influence on the temperature and humidity of North India, especially that of Jammu and Kashmir. By reason of its altitude and its situation directly in the path of the monsoons, it is most favourably conditioned for the precipitation of much of their contained moisture, either as rain or snow. Glaciers of enormous magnitude are nourished on higher ranges by this precipitation, which together with the abundant rainfall of the lower ranges, feeds a number of rivers, which course down to the plains in hundreds of fertilising streams. In this way, the Himalayas protect the Indian subcontinent from the gradual desiccation which overspreading the Central Asian continent, from Tibet northwards, and the desert conditions that inevitably follow continental desiccation. In the absence of the Himalayas, the climate, soils and natural vegetation of the Indian subcontinent would have been altogether different. The history of the peopling of the region and the mode of life of the people would also have been much different in the subcontinent in the absence of the Himalayas.

Glaciers of Jammu and Kashmir

The Great Himalayas or the inner most line of ranges of high altitudes reaching beyond 6,000 m are the enormous gathering grounds of snow which feed a multitude of glaciers, some of which are among the largest in the world outside the polar regions.

The majority of the Himalayan glaciers are from 3 to 5 km in length, but there are some giant glaciers of 40 km. The largest glaciers are those of the Karakoram, discharging into the Indus. The largest glaciers are the Hispar and the Batura of the Hunza Valley, 58 to 61 km long, while the Biafo and the Baltaro glaciers of the Shigar tributary of the Indus are about 69 km in length; the thickness of the ice-stream in these glaciers varies from 120 to 305 m. Still more mighty examples are the Siachin-glacier, falling into the Nubra tributary of Indus, some 72 km long. The thickness of the Zemu glacier is about 200 m while the Fedchenko (Pamir region) has a depth of nearly 550 m of ice. The southern side of the Great Himalayan chain has the glaciers of Biafo, Baltoro, Siachin, Remo and Braldu. Similarly, there are glaciers of enormous size in Kuen-Lun like Parkhchak in the Suru Valley (Kargil Dist). These glaciers are, however, dwindling at a faster pace.

The majority of the glaciers are of valley type. There are numerous hanging glaciers also. Some of the glaciers descend to such low levels as 2,100 to 2,500 m in the Kashmir Himalayas.[38]

Drainage System

The arrangement and disposition of streams which a drainage system etches into the land surface and which may reflect the sum total of factors which influence the number, size and frequency of streams in a particular area is known as a drainage system. The patterns of streams are influenced by: (i) initial slope, (ii) lithology, (iii) structure, (iv) the geological and geomorphological history of the area, and (v) the climate and rainfall regime of the area.

The drainage system of the Himalayas in general and that of the western Himalayas in particular is of much more recent origin owing to the mountain building movement of the late Tertiary Age. The most important fact about the rivers of the western Himalayas is that they are not consequent, i.e., their formation was not consequent upon the physical features or the relief of the region. In other words, many of the Himalayan rivers are older than the mountains they traverse. During the slow process of mountain formation by the folding, contortion and upheaval of the rock-beds, the old rivers kept very much to their own channels. The great momentum acquired by this upheaval was expanded in eroding their channels at a faster pace. Thus, the elevation of the mountain and the erosion of the valley—the two processes keeping pace with one another resulted into a completely developed valley system, intersecting it in deep transverse gorges or canons. Thus, the Himalayan drainage is *antecedent* in nature in which the main channels of river water were in existence before the present features of the region were impressed on it[39].

The state of Jammu and Kashmir is drained by the mighty Indus and its tributaries like, Kishan-Ganga, Jhelum, Chenab and Ravi and their tributaries. Out of these, the Indus and the Chenab rivers have their origins well to the north of the Greater Himalayas and they pierce through the main ranges of the Himalayas. The Jhelum has its origin at the Verinag near the Banihal-Tunnel in the Pir Panjal Range.

The Indus

Rising in the vicinity of Mansarowar Lake from Sengge-Khambal Glacier, the Indus river flows in a north-westerly direction through the trough between the Kailash and the Ladakh Range. It is 709 km, long within India and drains an area of about 117,844 sq km. For about 320 km it flows in a north-westerly direction, when it crosses on the south-eastern boundary of Kashmir at an elevation of about 4, 500 m. Flowing in the same direction, it falls to 3,500 m at Leh. About 19 km downward, it is joined by the Shyok river on the right bank. Near Skardu, the Shigar joins the Indus on the right bank. Giligt is another important right-hand tributary of the Indus. The river, thence, flows west, crosses the Kashmir border and turns south and south-west wards entering into Pakistan.

The river Indus though flowing through gorege has numerous alluvial fans and river terraces. The town of Leh is situated on an alluvial fan. To the east of Leh, on either bank of the Indus is an alluvial fan stretching over a distance of over 30 km. This is the most fertile and levelled land sprinkled with several rural settlements of the Ladakh region.

The flows in Indus are subject to extreme variations. The maximum flow of water in the river is in the summer months (June to August) when the snow melts. The low winter flows are mainly from the ground storage built up during summer. Thus a drought in winter may follow a flood in summer.

The Jhelum

The Hydaspes of the ancients (Greeks and Romans), the Vedasta of the Hindus, the Jhelum is known to the Kashmiris as the Veth. When it leaves Kashmir at Baramulla it is called the Kashur Darya and after joining the Kishnganga it is spoken as the Jhelum-river. Though Verinag is considered as its source, the local people maintain that a spring, a little below Verinag known as Vethvatru, is the source of Jhelum[40]. The river of Jhelum rises from the Pir Panjal Ranges about one km ahead of Verinag. Initially the river flows in a north-westerly direction and after passing through the city of Srinagar merges into the Wular Lake. Emerging from the Wular, it takes a south-westerly direction which it pursues upto Baramulla. It finally passes into Pakistan through the Baramulla-Uri gorge. The Jhelum flood plain is about 1,585m above the sea level.

The Jhelum river is navigable from Khanabal to Baramulla, a distance of about 170 km (106 miles). The great Wular Lake may be regarded as the delta of the Jhelum in Kashmir. In December, when the river is at its lowest, the average breadth is about 35 m and its mean depth only about 3 m.

On the right bank above Khanabal, the Jhelum river is joined by the Sandrin, the Bring, the Kokarnag, and the Achabal streams, and just below Khanabal on its right bank the Jhelum receives one of the most important tributaries, the Liddar, also known as Lambodri which has its source from the Tarsar Lake. The Sind Nala, the most important of all the tributaries of the Jhelum which joins Jhelum at Shadipur village (the place of confluence of two rivers) and after Wular Lake the Jhelum receives only one more tributary (Pohru) on its right bank before it reaches Baramulla. The Pohru river drains the Lalab Valley and joins the main river at Dubgam. The Erin and Madmati are also right-hand tributaries of the Jhelum (Fig.1.6).

The chief tributaries on the left bank are the Vishave, the Rambiara, the Ramshi and the Dudganga. The other left bank tributaries are the Suknag and the Firozpur Nala which lose themselves in the large marshes under the banks of the Jhelum. Of these small rivers, the Pohru, Sind and Visha are navigable for a short distance.

In greater part of the year, especially in autumn and winter seasons, the river flows gently but in times of flood it inundates and submerges the adjacent areas. At the occurrence of floods, great damage is caused to the crops of maize, linseed, barley, oats, rapeseed and wheat. The loss caused by floods is always more severe below the city of Srinagar.

In India, the river Jhelum flows for about 400 km and it drains about 28,500 km. The maximum and the minimum discharges of the Jhelum river recorded near the Indo-Pak border are as under:

Site	*Maximum Discharge*	*Minimum Discharge*
Near Indo-Pak Border	29,600 m^3	112 m^3

The Kishanganga

The river of Kishanganga is an important tributary of the Jhelum river. It has its origin in Kishansar Lake and consequently been named as Kishanganga. It drains the Tilail, Burzha-Bal, Gurez and their surrounding region. Kishanganga merges into the Jhelum river near the city of Muzaffarabad.

The Chenab

The Chenab is the mightiest river of the state. It has its source in the glaciers of the Middle Himalayas. At the source, the river is in two streams—the

Chandra and the Bhaga—at an elevation of about 4,900 m in Himachal Pradesh. After the confluence, the river flows between the Himalayas and the Pir Panjal Range, and then takes a turn to the south through a gorge, crossing Pir Panjal. It flows between mountain spurs till it enters the plain near Akhnoor. The Chenab is about 1,180 km long in the state of Jammu and Kashmir and its basin sprawls our 26,755 sq km. The maximum and the minimum discharge of the Chenab has been given below:

Site	*Maximum Discharge*	*Minimum Discharge*
Near the Actual Line of Control	15,400 m^3	102 m^3

The rivers of the state have enormous potential for the generation of hydel power. Unfortunately, these rivers have not been adequately harnessed. Inadequacy of funds, political unstability and lack of will are the main factors responsible for the underutilization of water potential of the rivers in the state.

Lakes

A natural depression filled with water and surrounded on all sides by land is known as a lake. It is an enclosed body of water usually, but not necessarily fresh water, from which the sea is excluded[41].

The state of Jammu and Kashmir is characterised by numerous lakes. The origin of these lakes may be attributed to the damming up of river valley (lakes in Ladakh), earth movements, landslips, glaciers, inundated hallows and the erosional hallows. The Wular, the Dal, the Anchar and the Manasbal are some of the most beautiful lakes of the world. These lakes are surrounded by the steep slopes of the mountains.

The Wular Lake

Most probably, the Wular has its origin from the Sanskrit world 'Ullola' meaning the lake with high waves[42]. In Kashmiri language Wular means 'cave'. It is said that at the place of the lake there was a city which was destroyed by an earthquake.

Wular is the largest fresh water lake in India, being about 80 sq km in area. It is almost surrounded by the lofty mountains which tower over the north and north-east of the valley. The Bohnar, Madhumati, and Erin rivers discharge their water into the lake, while from the south the Jhelum seeks a passage through the Wular to Baramulla. The lands around the Wular are never safe when the floods come down and a torrential rain in the surrounding areas with melting snow will spread the lake over several kilometres of the country. In the north-east corner is an island made by the great Kashmir king Zainul-Abaidin, and the ruins on it show that it must have been a place of great beauty. The lake is shrinking at a faster pace and the coastal shallow waters have been reclaimed for cultivation.

Dal Lake

Situated at an elevation of about 1580 m above the sea level the Dal Lake lies between 25^0 5′ north latitude and 74^0 51′ east longitude. The Dal has a backdrop of high Zaberwan Mountains (Fig.1.8). The surrounding alluvial lands are devoted to paddy, maize, vegetables and orchards.

The Dal having a catchment of about 316 sq km is divided into several distinct basins, namely, Gagribal, the Bod-Dal (large lake), the Astrawhol (Hazrat Bal Basin) and Dal Kotwal (Dudder Pokkar).

The lake is crossed by a narrow path running along a raised causeway called the Suttu or Sut-i-Chodri. This causeway starts from near the end of the Naid Yar Bridge in Karaliyar, and crossing the lake in a north-easterly direction terminates on the south side of the village Ishiburi, near Nishat Bagh (Fig.1.8).

Apart from precipitation (rainfall and snowing) the major sources of water of the Dal lake are : (i) the Telbal Nallah and its catchment, (ii) a numbers of small streams around the shores of the lake bringing outwash from the surrounding areas, and (iii) numerous springs having their sources at the bottom of the Dal. The area distribution, the open water and marshy lands have been shown in Fig. 1.8 (Table 1.1).

Table 1.1

Open Water and Marshy lands in the Various Components of Dal Lake-1991
(Area in sq. km.)

	Division	*Open Water*	*Marshy Land*	*Total Area*
1.	The Astawhol (HazratBal) Basin	5.6	3.2	8.8
2.	The Bod-Dal Basin	4.2	—	4.2
3.	Gagribal Basin	1.3	—	1.3
4.	Boulevard Basin	0.3	0.2	0.5
5.	Floating Gardens	0.3	4.5	4.8
	Total	11.7	7.9	19.6

Source: (1) After B.L. Koul.

It will be seen from Table 1.1 that out of the 19.6 sq km or of the total area of the Dal Lake only 11.7 sq km or nearly 60 per cent is under open water and the remaining 7.9 sq km (40%) is occupied by marshes and *demb* fields. In the open water also, there is less than 8 sq km area which has an av-

FIGURE 1.8

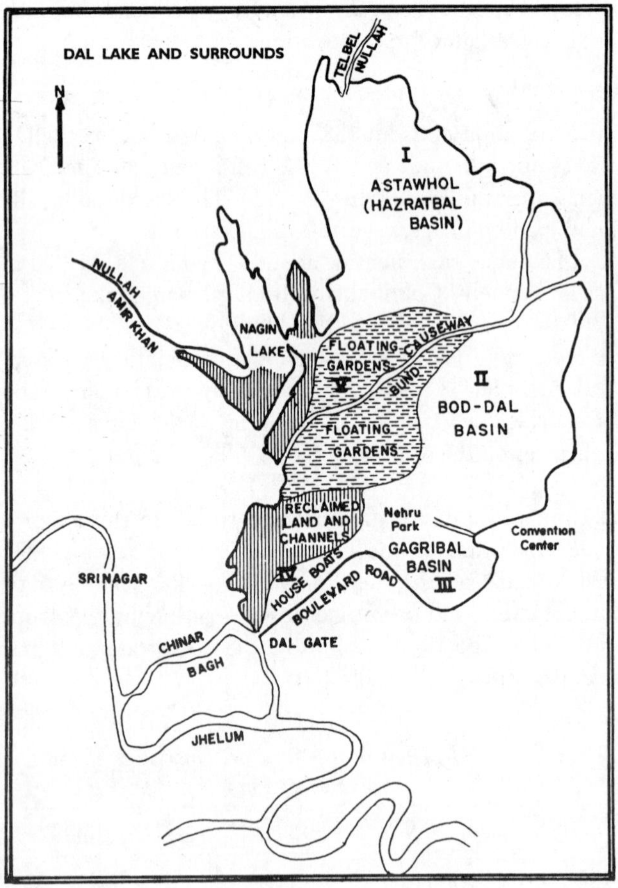

erage depth of more than two metres. The lake is situated in continental climate, characterised with marked seasonality. The seasonal change in water temperature indicates an almost homotherming setting in late October which continues upto April. Thereafter due to increasing water temperature a thermal gradient suggestive of thermal stratification develops.[43]

Changing Shoreline and Reduction of Open Water Area

The historical records, memoirs, and the travellers accounts show that the Dal Lake was quite large in size in the past. But after the 12th century its shoreline is continuously changing, thereby reducing its area and shallowing in depth. On the basis of the available historical records the areal extension of the Dal Lake has been given in Table 1.2.

Table 1.2

Dal Lake : Changing Area and Dimensions

Period	Source Name		Dimensions Length Breadth	Open Water in sq km
1. 6th Century A.D.	Nilmata Puran	Sureshwari	—	—
2. 12th Century A.D.	Kalhana's Rajaringini	Sureshwari	—	—
3. 15th Century A.D.	Srivara's Rajavali-Pataka	Sureshwari	19x4 km	76.00
4. 19th Century	Lawrence (1895)	Dal Lake	6.4x4 km	25.60
5. 20th C. (1930)	Survey of India	Dal Lake	5x4 km	20.00
6. 1981	Survey of India	Dal Lake	3.98 km x 2.94	11.70

Source: Historical Records (Rajtarangani, Ain-e-Akbari, Lawrence).

An examination of Table 1.2 vividly shows that during the Medieval Period, the area of the Dal Lake was 76 sq km. According to Lawrance its area around the ending part of the 19th century was over 25 sq km which got reduced to only 20 sq km in 1930. According to the Survey of India maps, in 1961 its area further declined and the open area covered only about 13 sq km. It was during the 20th century when most of the houseboats were anchored and a large number of *shikaras* were made for the tourists. In 1971, the open water area of the lake shrunk to 12 sq km with a further decline in its size in 1981 when its open water covered only about 11.70 sq km. It is not only that the size of the lake is shrinking and its shoreline is continuously changing, the rates of reduction in area and shallowing have been substantially accelerated.

The main causes of reduction in size of the Dal and its shallowness may be attributed to the following factors.

Inflow of Silt and Salinisation

As stated in the preceding paragraphs, there are numerous streams on all sides of the lake which discharge their inoragnic matter and nutrients into the Dal lake. Out of these, the Telbal Nallah which drains the deforested and overgrazed catchment, brings enormous quantities of silt and nutrients annually into the Astawhol Basin (Fig.1.8). Moreover, the chemical fertilizers, generally applied in the paddy fields, at the occurrence of heavy rains overflow into the lake and make its water more fertile. It has been estimated that the average annual accumulation of nitrate inflow in the lake is over 16 tonnes of phosphorous and 325 tonnes of nitrogen. In addition to this, about 21 tonnes of phosphorous and 145 tonnes of nitrogen are added to the weeds annually by other sources. Consequently, there is tremendous growth of macrophytes, phytoplanktons, zooplanktons, fish and bacteria in the lake.

During the last few decades, the adoption of High Yielding Varieties of rice and the application of NPK, insecticides and pesticides, the water of Dal Lake has become more saline. The pH value of water varies between 7.7 in the Bod Dal to 9.5 in the Astawhol Basin. The high pH value of water is however, detrimental to the aquatic life.

The second most important factor responsible for the degradation of the lake is the unplanned growth of the Srinagar City. Unfortunately, the city does not have an elaborate sewage disposal system.

Floating gardens occupy an area of about five sq km of the lake. These floating gardens are locally known as *Radh.* They provide fresh vegetables to the population of Srinagar but the huge quantity of garbage and waste produced by them is highly deliterious to the health of lake ecosystem.

The Dal Lake is a big attraction for the tourists, who prefer to stay in houseboats. There are around 400 housesboats of different categories and over 350 *Dongas* (residential boats). The sewage originating from these houseboats and *Dongas* is not treated before it is discharged into the Dal.

Deforestation and overgrazing on the slopes of Zaberwan have enhanced the process of soil erosion. Poverty and illiteracy of the lake-dwellers are also contributing to the depletion of Dal Lake.

Undoubtedly, the problem of Dal Lake environmental degradation has assumed a critical dimension as a result of which there is an undeniable need for environmental planning. If immediate remedial steps are not taken, the present generation would be condemned by the future generations for the careless management of the lake.

Mansbal Lake

This lake is about 2 km from the left bank of the Jhelum. The lake is about 5 km long, about 2 km wide and nearly 16 m deep. The lake is connected with the main river through a canal which enters it about 350 yards south of the village Sumbal. In addition to these, there are some mountain lakes such as Kousar-Nag, Shish-Nag, Ganga-Bal, and Sarbal-Nag.

In the Jammu Division also there are several important lakes, out of which Mansar and Sureinsar are more important.

Mansar Lake

This lake is situated near Parmandal and is about 15 km from the Sureinsar lake. It is about 1200 m in length and about 800 m in breadth. Its location is about 700 m above the sea level which is fed by natural springs in its bottom.

Sureinsar (Saroin-sar)

It is another most attractive lake of the Jammu Division. It is about 800 m in length and about 500 m in breadth. It is also fed by a springs and has a little island at the centre.

Springs

The state of Jammu and Kashmir is characterized with numerous springs. The people recognize the springs of great sanctity by the fact that their water in cold is summer and warm in winter. The springs are useful auxiliaries to the mountain-streams in irrigation, and are sometimes the sole source of water, as in the case of Achabal, Verinag, and Kokarnag on the south and Arpal in the east. Similar streams are found in the Jammu Division, while Ladakh Division has numerous warm water sulphur springs.

One of the most important and attractive springs is that of Achabal which gushes out of the Sosanwar Hill. It is said that the Brang river which disappears at Dewalgam in the limestone topography is the real source of Achabal spring. Verinag, a few km from the Banihal Tunnel is another important spring of great importance and beauty, with deep blue water also issues from the bottom of a high scaro of mountain spur. Anantnag (Islamabad), the place of countless springs, sends out numerous streams. One of the springs, the Maliknag is sulphurous and its water is highly prized for garden cultivation. All these springs are full of expensive trout fish. Chashma-Shali above the Dal Lake near Raj Bhavan is another famous springs. The rich people of Srinagar city procure their water, especially during the months of June, July and August from Chashma-Shali. The water of Chashma-Shali is considered to be the most clean and highly digestive which remedies indigestion.

Earthquakes

The state of Jammu and Kashmir falls within the great earthquake belt which traverses the Old Continents (Asia, Africa, Europe) from west to east. The state being the zone of weakness and largely occupied by the young folded mountains is subjected by the severe crumpling of the rock beds, resulting into earthquakes of various intensity and magnitude. It is well authenticated generalization that most of the Indian earthquakes including those which occurred in Kashmir and Pakistan have originated from the Great Plains of India or from the Peripheral Mountain Ranges to their north, west and east[44]. A few very destructive earthquakes have occurred in the state of Jammu and Kashmir in the last two hundred years.

According to the recorded data and information, the earthquake which occurred in April, 1873 took nearly 3,000 lives and a few thousand of cattle heads. Another earthquake was recorded on May 30, at 3 a.m. It was quite violent with rumbling sounds. A fissure about 225 m long, about 100 m wide and about 22 m deep occurred at Laridura in Kruhen Pargana to the south of Baramulla town. The effects of this earthquake were quite disastrous. Over 10,000 houses were destroyed and 3,400 human lives and 40,000 cattle heads were lost[45]. The shocks continued for two and a half months. Many irrigation springs disappeared and a big landslide occurred on the sloping ground of Laridura.

The most disastrous earthquake occurred on 30 May, 1885 at 3 a.m. This earthquake was felt over an area of about 110,000 sq miles. Its epicentre was a few miles to the west of Srinagar. Subsequently, shocks were felt upto 16 August. Thousands of houses were destroyed and there was an atmosphere of general panic. People slept for many days in the open. About 3,500 persons were killed and the number of cattle heads, ponies, goats, poultry, etc., was enormous which were crushed by falling buildings and earthen houses.

Baramulla and Pattan towns suffered most and large fissures were caused from which sulphur fumes and inflammable gases were emitted. Many old water springs disappeared and numerous landslides occurred.

Another earthquake occurred on 20 September, 1902. The epicentre of this earthquake was near Hindukush and it was felt all over the Kashmir Valley. At Srinagar, it lasted for about two and a half minutes. It was also felt severely at Gulmarg and Drosh. The loss to life and property in this earthquake was insignificant.

A disastrous earthquake occurred on 4 April, 1905. Its magnitude at the Richter Scale was 8. The first shock was recorded at 6.10 a.m. in Srinagar and it lasted for about three and a half minutes. It was also recorded at Leh, Kargil, Skardu, Uri,Dras and Muzaffarabad. The shocks were of mild intensity and much damage was not done to life and property. Earthquakes of mild intensity were also recorded in 1907, 1912, 1917, 1921, 1931, 1952, 1984, 1986, 1992, 1998, and 1999.

So far as the Jammu Province is concerned, the whole of it experienced a severe shock in 1905 which caused a few deaths and considerable damage to property. The Ladakh Province, however, recorded frequent shocks of mild intensity.

Notes

1. D.N. Wadia, 1994, *Geology of India,* Tata MaGraw-Hill, 4th ed. p. 73.
2. C.A. McMahon, Microscopic Structure of some Himalayan Granites *Rec.*

G.S.I. vol. xvii. pt. 2,1984.

3. D.N. Wadia, 1994, op. cit., pp. 87-88.
4. Sir T.H. Holland at G.H. Tipper *Mem.G.S.I.* xLiii, pt.I, 1913, and (Second Ed.) *Mem li.* pt. I, 1926, The Dharwar System.
5. C.S. Middlemiss, Silurian-Trias Sequence of South-East Kashmir, *Rec.G.S.I.,* vol.xL,pt. 3,1910.
6. D.N. Wadia, op. cit., pp. 153-154.
7. Ibid. p. 207.
8. D.N. Wadia, Cambrian-Trias Sequence of N.W. Kashmir, *Rec. G.S.I.* vol. 1xviii, pt. 2, 1934.
9. Karl Diener, Trias of the Himalayas, *Mem. G.S.I.* vol. xxxvi, pt. 3, 1912.
10. D.N. Wadia, op. cit., pp. 243-44.
11. D.G.A. Whitten and J.R.V. Brooks, *The Penguin Dictionary of Geology,* 1977.
12. Arnold Heim, *Himalayan Journal,* ix. p. 41, 1937.
13. D.N. Wadia, *Rec. G.S.I.* vol. xxii, pt. 2, 1937.
14. H.B. Medlicott, *Geological Structure and Relations of the Himalayas,* Mem. G.S.I. vol. iii pt. 2, 1864.
15. Ibid.
16. E.S. Pinfol, Structure and Stratigraply of N.W. Punjab, *Rec. G.S.I.* vol. xix, pt. 3, 1918.
17. W.T. Blandford, Geology of Western Sind *Mem. G.S.I.* vol. xvii pt. I, 1879.
18. C.S. Middlemiss, Geology of the Sub-Himalayas, *Mem. G.S.I.* vol.xxiv. pt. 2,1890.
19. D.N. Wadia, op. cit. p. 346.
20. D.N. Wadia, Siwaliks of Potwar and Jammu Hills, *Mem. G.S.I.* vol. 1i pt. 2, 1928.
21. D.N. Wadia, Pleistocene Ice Age Deposits of Kashmir, Proc. *Nat. Inst. Sc. India 7,* 1941.
22. Ibid.
23. H.de Terra, Studies on the Ice Age Deposits of Kashmir, *Proc. Nat. Inst. Sc. India 7,* 1941.
24. Ibid.
25. F. Drew, *Jammu and Kashmir Territories,* 1875.
26. E.A. Glennie, Gravity Anomalies and Structure of the Earth Crust, *Survey of India,* Dehra Dun, 1932.
27. M. Husain, *Geography of Jammu and Kashmir State,* 1986, Rajesh Pub. New Delhi-2.
28. Ibid. p.5.
29. C.S. Middlemiss, *Rec. G.S.I.* vol. XLI, pt. 2,1911.
30. D.N.Wadia, Geology of Poonch and Adjoining Area, *Mem. G.S.I.* vol. Li, pt. 2, 1928.

31. W.R. Lawrance, *The Valley of Kashmir*, Navbharat Printing Press, Delhi, 1996.
32. Ibid. p.44.
33. Ibid.
34. Ibid. p. 45.
35. D.N. Wadia, op. cit. p.13.
36. Ibid. p.14.
37. Ibid. p. 1.
38. J.S. Dev, 1983, *Natural Calamities of Jammu and Kashmir,* New Delhi, Ariana Pub. pp. 95-105.
39. D.N. Wadia, op. cit. p. 28.
40. Lawrance, W.R. op. cit. p. 18.
41. Whittow, J., 1984, *The Penguin Dictionary of Physical Geography,* p. 298.
42. Lawrance, W.R. op. cit. p. 21.
43. Koul, B.L. 1964, "Shrinkage of Dal Lake" in *Sustainable Development and Environmental Impact Assessment*, ed. M. Husain, Rima Pub. New Delhi.
44. J.S. Dev, op. cit.
45. Ibid.

2

Climate

Climate has been defined as the average weather conditions at a specific place over a lengthy period of time, i.e., more than 30 years, while the conditions of the atmosphere at any place at a specific time or for a short time is known as weather. In modern usage, the term is used most commonly of world regions other than of particular places. Climate deals with all the meteorological elements, i.e., atmospheric pressure, humidity, precipitation, temperature, wind and the way they are influenced by latitude, altitude and aspect of slope, etc.

Situated in the subtropical latitudes and having micro-level variation is altitudes, the state of Jammu and Kashmir has great diversity in its temperature and precipitation. Excepting the plain, south of the Siwaliks of the Jammu Division, the climate over the greater parts of the state resembles to the mountainous and continental parts of the temperate latitudes. In the Koppen's scheme of climatic types, the Ladakh Division has been shown as the Dwd (extremely cold and dry winter and short cool summer, and the hilly parts of Jammu and Kashmir Divisions as Dfb (humid continental with severe and moist winter and short summers); while the plain areas of the Jammu Division, adjacent to the plains of Punjab have been described as Ca (less most winter, long hot summers, followed by moderately heavy rains in the months of July, August and September[1].

As discussed in the preceding chapter, the state of Jammu and Kashmir is essentially mountainous, characterised by undulating relief of steep slopes, deep gorges, snow-covered peaks and rugged topography. Its altitudinal variations vary between 274 m in the plain of Jammu to 8,615 m-Mt. K[2] in Karakoram (Ladakh), and Nanga-Parbat (8119 m) in Kashmir Himalayas. The normal lapse rate of temperature entails an average decrease of about 6°C per thousand metres increase of elevation. There are important exceptions to this is actual lapse rates, but on the whole, higher altitudes are associated with significantly lower temperatures and smaller annual range of temperature. Air pressure always decreases with altitude, the atmosphere above areas at high elevations is ordinarily much freer of dust, smoke, and other non-gaseous materials, and the air is accordingly more transparent to the passage of both incoming and outgoing radiation during cloudless period. Temperatures and precipitation vary from place to place, from slope to slope and from windward to the leeward side. Still the number of meteorological stations in the mountainous regions of the state are very few. The rainfall recording stations are inadequate to describe and interpret the micro-level variations in the general weather and climatic conditions of the state. Looking at the highly diversified terrain and latitudinal extension of the state it would be appropriate to discuss the climate of the Jammu, Kashmir and Ladakh divisions separately.

Climate of the Jammu Division

From the point of view of climate, the Jammu Division is divisible into two parts, i.e., (i) the plain region, lying to the south of the Siwaliks, and (ii) the mountainous region, stretching over the Middle and the Greater Himalayas in the districts of Doda, Rajauri, Poonch and Udhampur.

The climate of the plain region and the Middle Himalayas including the Pir Panjal in characterised by a rhythm of seasons which is caused by the reversal of winds in the form of the south-west and the north-east monsoons. The reversal of pressure takes place regularly twice in the course of a year. The mean monthly temperature at the Jammu City reads well above the 20°C, the annual range of temperature being about 17°C. The outstanding feature of the annual march of temperature is the occurrence of maximum before the commencement of summer monsoon, i.e., in the month of May or June. There may be a secondary maximum again just after the rainy season in the second half of September. In general, the maximum temperatures occur during the period of clear skies, increasing isolation and a water-budget deficit before the onset of more persistent cloudiness and heavy rainfall (Fig.2.1).

In the plain and lower reaches of the Himalayas, the temperature starts rising in the month of March. The mean maximum and mean minimum temperatures in March being 23°C and 12°C at Jammu. In the month of April the day and night temperatures read 32°C and 18°C respectively which jumps to 38°C and 25°C in the month of May. June is the hottest month at Jammu, recording a mean monthly temperature of about 31.65°C. On individual day in the month of May or June the maximum temperature, however, may read upto 47°C (June 1953). The relative humidity in May falls to below 20 per cent. All the towns lying to the south of the Siwaliks (Akhnur, Hiranagar, Kathua, Samba, etc.) observe intense tropical heat in the months of May and June. During the summer season in the City of Jammu and its environs, a cool wind descends from the Siwaliks in the night

FIGURE 2.1

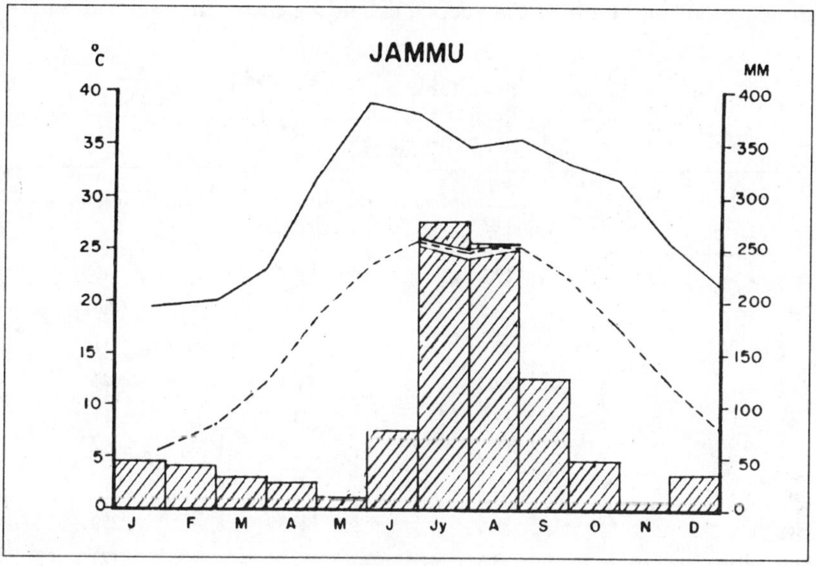

FIGURE 2.2

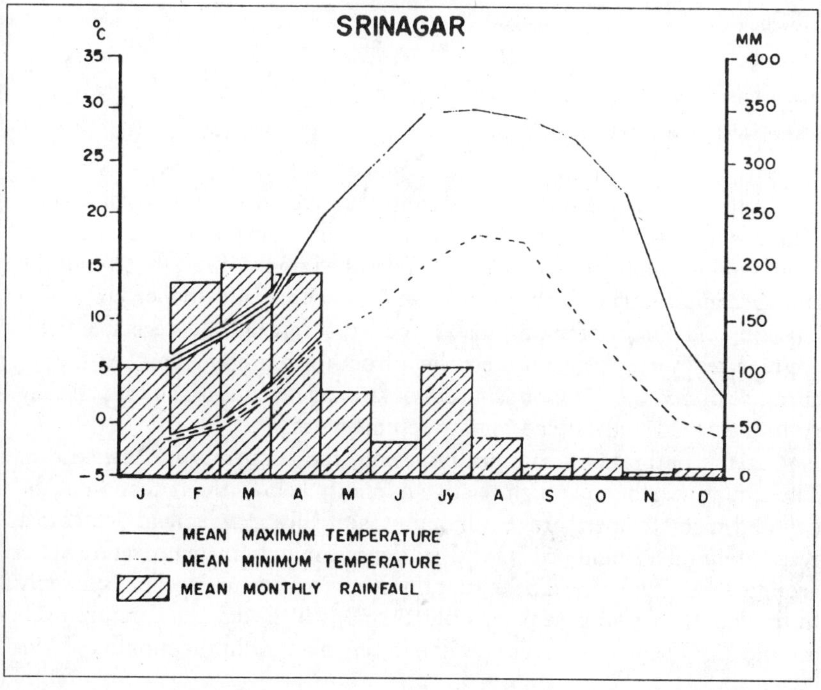

time which is locally known as *Dadu*. This cool local wind has a salubrious influence on the health of the people, enhancing their efficiency and vitality.[2]

Table 2.1

Jammu: Mean Maximum, Mean Minimum Temperatures and Mean Monthly Rainfall (1965-1999)

	Mean Maximum Temp.°C	*Mean Minimum Temp.°C*	*Rainfall in mm.*
January	19.0	6.7	46
February	20.2	8.4	40
March	23.4	12.0	31
April	32.2	18.1	24
May	38.3	23.0	11
June	38.0	25.3	75
July	34.2	25.1	275
August	35.5	24.8	250
September	33.6	22.4	125
October	31.9	17.5	46
November	26.4	12.3	08
December	22.2	7.7	33
Mean/total	29.39	17.19	96.34

Source: Meteorological Deptt. Srinagar.

In the summer season the mild morning breeze becomes warm around 10 a.m. and hot in the noon which blows steadily from west to east. This hot local wind is locally known as *'Loo'*. Under the impact of *Loo* (hot wind) the land becomes parched, the green vegetables wither and fodder crops like *Barseem*, and *Ruzka* orchards, maize, fodder, sugarcane and vegetable fields are repeatedly irrigated to protect them from the scorching heat. At the occurrence of *Loo* the outdoor activities of farmers remain suspended and they go to their fields only after 4 p.m. to resume work.

At the outbreak of summer monsoon, the temperature comes down. The cloudy weather and high relative humidity help in the reduction of the day and night temperatures in the months of July, August and September read around 34°C and 35°C, while the mean minimum temperature reads around 25°C (Table 2.1).The mean minimum temperature declines suddenly in the month of October, being only 17.5°C, yet the day temperature reads around 32°C. There is further decline in the mercury in the months of November. The winters are well set in December in which the mean maximum and mean minimum temperature comes down to 22°C and 7.7°C respectively. January is the coldest month in the plain areas of Jammu in which the average monthly temperature reads around 13°C (Table 2.1).

The occurrence of frost, fog and mist is quite frequent in the morning hours of December, January and February. At the occurrence of fog and mist the visibility is reduced to few metres which hampers the movement of vehicular traffic and the take off and landing of the aeroplanes. The frost is highly injurious to vegetables, orchards, floriculture and sugarcane. The chillies wither immediately at the occurrence of frost. Moreover, in the cold weather season cold wind blows from north to south, i.e., from the snow-covered peaks and ranges of the greater Himalayas to the plain areas. The biting cold winds and the fluctuating relative humidity makes the weather either keen or raw. The Siwaliks and the Nawanchoch ridge in the vicinity of the plain act as a barrier against the cold and dry winds from the north.

The climate and weather conditions of the mountainous tract of the Jammu Division are altogether different from that of the plain areas. In the mountainous areas, the Pir Panjal Range of the Lower Himalayas, the snow-covered peaks of the Greater Himalayas, the rugged topography and aspect of slope have a close influence on the temperature and precipitation. The orientation of the valleys, whether they are longitudinal or transverse, also influence significantly the general weather conditions. It is because of these factors that Ramban town, situated on the right bank of Chenab and surrounded by barren, dark coloured rocks, records higher temperatures to that of Batote, though both of them are located within a crow-fly distance of about 25 km from each other.

The Upper Chenab region stretches over the Greater Himalayas. The meteorological stations of Bhadarwah and Kishtwar situated in this region record very low temperatures (-15°C) in the winter season. In the Lower and Upper Himalayas of the Jammu Division, although rainfall recording stations are at Banihal, Batote, Bhadarwah Doda, Kishtwar, Poonch, Rajauri Ramban, Ramsu, Riasi etc. have been established, reliable data on temperature, pressure and relative humidity are not available. Being situated in higher altitudes, the winters at all these stations are severe and they record light to heavy snowing in the month of December, January and February. The presence of lofty mountain ranges prevent the penetration of warm winds from the scorching plains of the south. The months of March to May are pleasantly warm in the mountainous region of the Jammu Division.

Rainfall in the hilly tract is recorded in almost all the twelve months of a year with a maxima in July and August. In fact, about 75 per cent of the overage annual rainfall at all the rainfall recording stations of the hilly and mountainous areas is recorded during the season of *Barsat* (general rains). At the occurrence of heavy rains in July and August and January and February there occur heavy landslides along the Jammu-Srinagar National Highway, especially in the *Nashri* area between Batote Rambam and Banihal.

The months of July and August are the wettest at Jammu, recording an average monthly rainfall of about 275 and 250 mm respectively. There is occasional thundering and lightning in the rainy and winter seasons, thundering activity in however, greatest at the beginning and the ending part of the rainy season. As the rainy season becomes established, the cumulous and nimbus clouds persist and bring more rainfall in July and August. The month of November is almost rainless in which the average monthly rainfall is only about 8 mm. It is the period of clear skies and cool morning breeze which is considered to be the month of invigorating climate.

The monsoon reaches the Jammu Division in the last week of June or in the first week of July and lasts till the 20 of September. The southern parts of the Jammu Division in which Akhnur, Basoli, Bishnah, Hiranagar, Kathua, Jammu and Samba are situated most of the rainfall comes in the season of advancing monsoon. As the altitude increases towards the Middle Himalayan mountains (Batote, Bhadarwah, Ramban, Banihal, Kishtwar, etc). The summer rainfall averages more, being around 1500 mm. The average annual rainfall at the Jammu city is about 965 mm out of which 725 mm or 75 per cent come during the season of general rains.

Rainfall in the winter season occurs from the western disturbances. The average rainfall at Jammu during winters (December to March) is about 150 mm. This rainfall is highly beneficial to the standing *rabi* crops of wheat, peas, gram, fodder and vegetables. In areas of higher altitudes of the Greater Himalayas and the Pir Panjal Range snowing is frequent from December to March. A marked characteristic of rainfall of the Jammu Division is the decrease in precipitation from east to west and north-west. Moreover, leeward slopes record less rainfall as compared to the windward slopes.

Climate of the Kashmir Division

The weather and climate of the Kashmir Division are intrinsically linked with the weather mechanism in the subcontinent in general. The location of the Kashmir Valley at a high altitude (about 1600 m) in the north-western corner of the subcontinent, surrounded by high mountains on all sides, give it a unique geographical character with distinctive climatic characteristics. In general, the distinctive features of the climate of Kashmir are: (i)mild summers, (ii) vigorous and severe winters with snowing and rain, (iii) a muggy and oppressive weather in July and August, and (iv) the most exquisitely pleasant spring.

In the Koppen's classification of climate the Valley of Kashmir may be placed in the Dfb category, having humid severe, raw winters and short mild summers. On the basis of temperature and precipitation a year in the Valley of Kashmir may be divided into the following four seasons:

1. Winter Season (November to February).
2. Spring Season (March to mid-May).
3. Summer Season (mid-May to mid-September).
4. Autumn Season (mid-September to October).

The mean maximum, and the mean minimum temperatures as well as the overage monthly rainfall of Srinagar have been given in Table 2.2 (Fig.2.2).

Table 2.2

Srinagar: Mean Maximum and Mean Minimum Temperatures and Mean Monthly Rainfall (1965-99)

	Mean Maximum Temp.°C	*Mean Minimum Temp.°C*	*Rainfall in mm*
January	5.8	-1.6	102
February	8.1	-0.5	183
March	12.1	3.00	202
April	20.1	8.1	180
May	24.9	11.2	86
June	29.4	15.0	35
July	29.6	18.2	113
August	28.9	17.5	37
September	27.1	10.8	13
October	22.1	5.6	15
November	14.4	1-5	08
December	8.2	-1.4	10
Mean/Total	19.27	7.29	984

Source: Meteorological Department, Srinagar.

Winter Season

In the Valley of Kashmir, the winter season lasts from November to February. The mean maximum and the mean minimum temperatures in November at Srinagar read about 14°C and 1.5°C respectively. The temperature substantially declines in December, the mean maximum and the mean minimum being about 8°C and 1.4°C respectively. In exceptional years of severe cold, e.g., the last week of December 1985, 1986, 1987, 1988 and 1994 in

Srinagar, the temperature went below -15°C. Consequently, most of the lakes of the valley floor (Dal, Wular, Anchar, Manasbal) were frozen and became the playgrounds for the adventurous children and sportsmen of Srinagar and its environs.

January is the coldest month in Kashmir which is locally known as the period of *Chilla-Kalan* (40 days of intense cold). Occurrence of heavy snow in this month in Kashmir is a common phenomenon. The non-occurrence of snow and rainfall in January is considered as a 'bad omen', leading to numerous epidemic disease and scarcity of water in the rivers and springs during the subsequent seasons. The winter season receives about 120 cm of snow, the maximum snowing being in the month of January, i.e. about 42 per cent of the total snowfall. The greater calamity in the winter is the occurrence of *Kot-Kushu* (frost) which seems to freeze the damp moisture of the air and destroys trees and even congeals eggs. Covered in by ice and snow with frozen lakes and freezing rivers, Kashmir during winters is like a huge refrigerator. In the winters there remains an acute shortage of fresh vegetables and at the occurrence of landslides in the Nashri area (between Batote and Ramban) the vehicular traffic between Jammu and Srinagar remains suspended for days or weeks together as a result of which the supply of sheep, goats and poultry gets disrupted. Consequently, vegetables and meat prices in the valley soar and the people of the valley face great difficulties in this period. Moreover, the *Balan* (fuelwood) price goes up, the outdoor activities remain suspended and life of the ordinary people becomes miserable.

In the winter season, the rains may be either partial or may be what the Kashmiri call *Alamgir* (widespread). It is believed that when the clouds gather over the Kausarnag Lake on the south-west mountains and over the Wular Lake in the north, the Kashmir can predict with some degree of certainty that there will be a general rain over the valley and 24 hours of rain will cause heavy floods in the river. The local adage based on the empirical experience of the peace about the different types of clouds and their occurrence at different parts of the day have been expressed in the following adage:

'Red in the night the shepherd's delight
Red in the morning the shepherd's warning.'

If the clouds (*Abras Hetun Nar*) have caught fire (become red in colour) in the evening, is a sure prelude to fine, sunny weather; *Nihadu*, which is red in the morning, presages (warning) rain. The Kashmiri folk have a strong conviction about the colour and nature of clouds. In their opinion white clouds (cirrus) are certain to bring heavy rain; dark clouds mean no rain or light showers.

In winters, rainfall occurs from the western disturbances (temperate cyclones). These disturbances have their origin in the Mediterranean Sea. The rainfall generated by these cyclones is fairly widespread locally known as *Alamgir.* There are, however, significant variations in the Kashmir Division. For example, the city of Srinagar records about 30 cm of rainfall during winter season (November to February), while rainfall during the corresponding period at Baramulla and Anantnag is about 35 and 34 cm respectively. Winter rainfall is much higher on either extremity of the valley. The town of Doru situated in the Anantnag District receives much higher quantities (45 cm) of rainfall to that of Srinagar.

October and November are usually dry months but in December the sky becomes overcast with cloud and haze, and by the third week of December the snow sets in.

In general, the Kashmir Division has highly dreary and monotonous winters with very little sunshine. The relative humidity usually remains around 90 per cent in winters. To combat cold people put on *Pheran* (a loose woolen garment), use *Kangri* (earthen firepot) and *Bukhari* (indigenous room-heater in which coal is burn to make the house warm). Under these conditions of severe cold, the consumption of *Balan* (fuelwood) is enormous. The common people have to spend a substantial proportion of their earnings in the purchase of fuelwood and coal. It is the period when there is an acute shortage of fresh vegetables and the supply of sheep, goats and vegetables from the Jammu Division, Punjab and Rajasthan is also disrupted, owing to landslides in the *Nashri* (National Highway between Batote, Ramban and Banihal). The agricultural activities remain suspended, leading to rural Unemployment. Consequently, many of the labourers and marginal farmers outmigrate from the valley to the different cities and towns of the Punjab, Haryana, Himachal Pradesh, Uttar Pradesh, Rajasthan and Delhi to find employment in assorted activities just to pass their winters away from the dreaded weather of the Kashmir Valley.

Spring Season

March to mid-May is the season of spring locally known as *Sonth.* At the advent of March the weather starts improving and the mercury goes up steadily. In March, the day temperature fluctuates between 10°C to 16°C. The mean minimum temperature remains around 3°C as a result of which nights are cold. There are, however, wide variations in the day and might temperatures at various altitudes. Although snow starts melting with the increase in temperature and many of the streams inundate, yet there occur one or two snowfalls in this month. In 1981 and 1982, there was heavy snowing in the last week of March. In fact, March is the wettest month, having a high frequency of western disturbances and recording about 200 mm average rain-

FIGURE 2.3

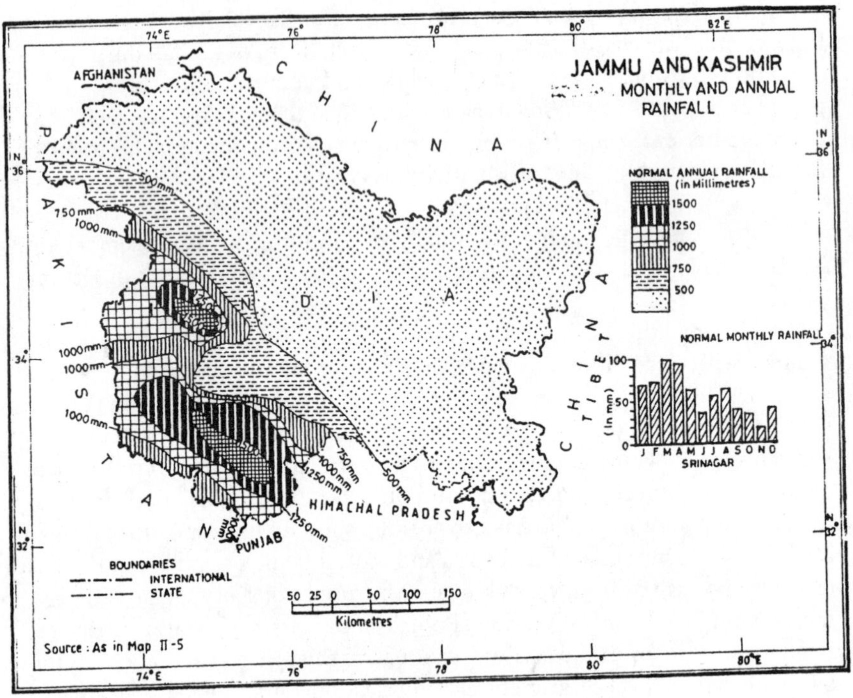

fall (Fig.2.3). The leaves on the deciduous trees like willow and *Chinar* (maple) appear in the last week of March or the first week of April.

The day temperature shoots up abruptly in the month of April. The night temperature also registers a steady increase, the mean maximum and the mean minimum in this month being about 21°C and 8°C respectively. there is a further increase in the average temperature of May, the day and night temperatures being about 25°C and 11°C respectively. There is a decline in the mean monthly rainfall. About 30 to 40 per cent of the total annual rainfall is recorded in the spring season at most of the rainfall recording stations. The rivers discharge large quantities of water as the melting of ice and snow in the high altitudes accelerates. This is the time when the transplantation of paddy takes place and fodder crops like maize, millets are sown.

In the spring season, with the steady increase in temperatures, lush green grasses develop in the *Margs* (alpine pastures). April and May are the months when the valley is in full bloom and the flowers of almond, apple, peach, pears and cherry (*Glass*) add more fragrance to the fresh and salubrious air of the valley. Agricultural activities which remain suspended during the long winters are again started in the ending parts of April and the first week of May. The seeds of vegetables are germinated in the house to be planted in the *Radhs* (floating gardens) of the Dal Lake. Ploughing of paddy fields and preparation of rice nurseries also commence in the early part of May.

The Summer Season

June to September is the period of summer season in the Valley of Kashmir. The mean monthly temperature in May at Srinagar reads about 22°C, the mean maximum and mean minimum being about 29.4°C and 15°C. July is the hottest month in which the maximum temperature on a particular day may shoot upto 35°C at Srinagar. The mean maximum and mean minimum temperatures in this month fluctuate between 30°C and 18°C respectively. Under the prevailing clear weather, blue skies and rarefied atmosphere this is the muggy, oppressive, unpleasant in which mosquitoes swarm. July and August are the months when electric fans are needed and the rich residents of Srinagar plan their outings to Gulmarg, Pahalgam and other neighbouring hill resorts which being at higher altitudes record relatively low temperatures and pleasant weather.

June to September in the period of summer monsoon in the Indian subcontinent. The Valley of Kashmir being surrounded by high mountains on all sides receives very little rainfall. For example, at Srinagar the total rainfall in this period (June to September) is only 51.3 mm which is only about 6 per cent of its average annual rainfall. The relative humidity in these months oscillates between 40 to 60 per cent. The sun in the months of June and July is bright, and seems as powerful just after sunrise and before sunset as it is at noon. The dryness of the air makes work in the valley very trying and tiresome. In fact, working in the paddy, maize and saffron fields after 10 a.m. becomes difficult. The elderly people of the valley opine that high temperatures in July and August is a recent phenomenon, and they attribute it to deforestation, high growth of population and the consequent increase in pollution. The occasional showers in the months of June, July and August have especial importance for the saffron growers which help them in the easy and smooth digging of saffron fields (karewa lands). Showers sometimes fall in the beginning of September. These showers are locally known as *Kambarka* which are considered of immense importance to the spring crops of the subsequent year.

In September, the temperature begins to fall, the nights become cool and by the end of this month the evenings are almost cold. The mean maximum and the mean minimum temperatures in this month at Srinagar read as 27°C and 11°C respectively. The average monthly rainfall in September is only 13 mm (Fig. 2.2).

Autumn Season

Mid-September to October is the period of autumn season in the Kashmir Division. These months mark a a transition from the warm subtropical summers to temperate winters. The months of autumn are characterised with the least disturbed weather. In this season, the skies generally remain clear, the duration of sunshine is longer and very little rainfall is recorded. October and November are the two months with bright, pleasant sun and cold night, and October is considered to be the healthiest and most delightful of all months in Kashmir. This season attracts the largest number of tourists in the valley from the various parts of the country and abroad.

In the month of October the mean maximum and the mean minimum temperatures read about 22°C and 5.6°C respectively. At the occurrence of clouds, the temperature, however, slumps abruptly and may read only 15°C in the day time and 2°C in the nights. The diurnal range of temperature is about 17°C. During the last sixty years, snowing in the month of October was recorded five times in the valley part of Kashmir.

The cool nights and warm days of September and October help in the ripening of walnut, almond, apples, and the later part of October when the night temperature reads about 2°C the weather is conducive for the emergence of saffron flowers. From the third week of October to the 15th of November is the harvesting period of saffron flowers. During autumn season the people collect and purchase *Balan* (fuelwood) for the ensuing winters and harvest the numerous varieties of apples. The main staple food crop (rice) is also harvested in the months of autumn season. After the main harvest of rice, saffron, apple and almonds, the Kashmir arrange the wedding of their sons and daughters. Thus, it is the period of great enjoyment and social interaction. The Kashmiris on the basis of their empirical experience divide a year into six seasons with a duration of two months each. The Kashmiri calendar of seasons is as under:

(i) *Sonth* (spring season), mid-March to mid-May.
(ii) *Grishm* (summer season) mid-May to mid-July.
(iii) *Wahrat* (rainy season) mid-July to mid September.
(iv) *Harud* (autumn season) mid-September to mid-November.
(v) *Wand* (winter season) mid-November to mid-January.
(vi) *Sheshur* (season of severe cold) mid-January to mid-March.

The general prevailing temperature and precipitation conditions of the given seasons have already been described in the preceding pages.

Climographs of the Jammu and Srinagar Cities

A general diagram to illustrate the general climatic character of the Jammu and Srinagar cities has been given in the form of climographs in Fig. 2.4. Cli-

FIGURE 2.4

CLIMOGRAPH
JAMMU & SRINAGAR

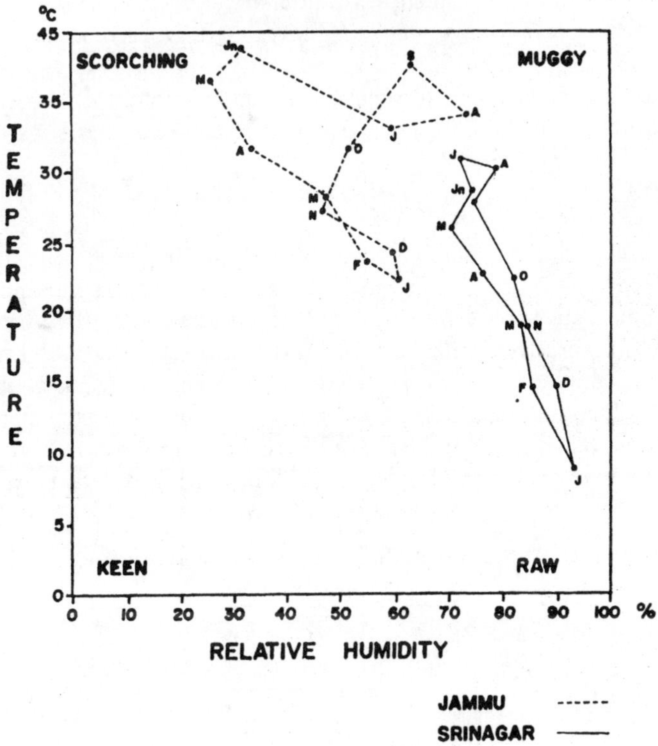

mograph provides an insight about the average temperature and relative humidity at a city/station on which depends human comfort.

It may be observed from Fig. 2.4 that in the Jammu City the mean monthly temperatures are fairly high from April to September and the relative humidity is only around 30 to 40 per cent between April to June. Consequently, the weather remains scorching during the summers season. In the rainy months (July-September,) the weather is muggy leading to uncomfortable oppressive conditions. The winters at Jammu are however, pleasant in which the mean monthly temperatures vary between 12°C to 18°C and relative humidity fluctuates between 50 to 70 per cent (Fig. 2.4).

Contrary to this, the climograph of Srinagar City shows that in the summer the weather conditions are slightly muggy, while the winters are extremely cold. The winters in Srinagar record a mean monthly temperatures around 4°C and the relative humidity between 70 to 90 per cent. Consequently, November to March, the weather is raw and highly uncomfortable. The people of Kashmir call such a cold and wet weather as the period of *Chilla-kalan*, *Chilla-Khaurd* and *Chilla-Bacha*. It is become of the raw weather and snowy conditions in winter that the outdoor activities remain suspended and people have to spent a substantial amount of their earnings on *Balan* (fuelwood and coal) to keep their bodies and houses warm (Fig. 2.4).

Climate of the Ladakh Division

Ladakh Division, lying mainly to the north of the Greater Himalayas, has a unique geographical personality. It is characterised with parallel mountain ranges, numerous snow-covered peaks, gigantic glaciers, narrow fertile valleys, alluvial fans, river terraces and seasonal lush green pastures. For most of the part Ladakh has bare crats, barren rocks and granite tablelands, devoid of natural vegetation. All these factors have closely influenced the climate of Ladakh. Moreover, the influence of local relief in most pronounced in respect of the effectiveness of isolation on slopes having different exposures

Table 2.3

Leh-Mean Maximum, Mean Minimum Temperatures and Mean Monthly Rainfall (1965-1999)

	Mean Max. Temp.°c	*Mean Min. Temp.°c*	*Average Monthly Rainfall mm*
January	0	-15.0	3.9
February	2.5	-13.9	3.2
March	6.5	-8.7	2.8
April	13.0	-4.0	2.3
May	16.5	0.0	2.2
June	21.9	3.2	1.3
July	24.5	7.5	4.5
August	23.0	5.6	5.5
September	18.8	1.3	2.8
October	12.0	-5.2	1.0
November	6.5	-8.5	0.3
December	2.8	-11.5	0.2
Mean/Total	12.33	-4.10	30.00

Source: Meteorological Department, New Delhi.

and with the modification of wind direction and speed. Every variation in slope with respect to the sun rays produces a different micro-climate. The prevailing climatic and weather conditions of Ladakh may be properly understood by making a systematic analysis of the mean monthly temperature and average monthly rainfall of the Leh Meteorological Station. The climate of Leh is, however, a representative of the limited area as there are significant micro-level variations in the climatic conditions of Ladakh Division.

In general the climate of Ladakh Division is very cold, arid and dry resembling to that of a cold desert. The winters are cold, dreary and depressing. The temperatures in winter months are extremely low in Daras, Rupsho. Tansktse, Rong and Zanskar, while central Ladakh and Nubra Valley have comparatively mild temperatures.

January is the coldest month in Ladakh. The lowest official surface temperature -40°C was resorted at Dras in 1986. But people of the Suru Valley on the basis of their empirical experience opine that the lowest temperatures in Ladakh are recorded at Rangdum, the unofficial lowest temperature being -50°C in January 1986. Under the calm conditions of night, accompanied with very low temperatures, the dry air does not produce the same physiological chilling as moisture air would at relatively higher temperatures. With wind, however, chilling effects rapidly increase and powdered snow may pervade the air which reduces she visibility significantly in the hilly areas. The long duration of cold and dry weather in the night may be appreciated from the fact that for a period of about nine months (September to May) the night temperature remains below the freezing point at Leh. In Ladakh the skies remain clear, the insolation is intense and the wind blows at a steady speed. Such atmospheric conditions are very conducive for the development of wind and solar energy.

The temperature increases rapidly in the month of April being 13°C in the day time while night temperature remains around -40°C. In the month of May, the mean maximum and the mean minimum temperatures read about 17°C and zero degree respectively. Between June to August the mean day temperatures fluctuate between 22° and 25°C but the mean minimum temperature remains below 8°C. Thus, nights are cold throughout the year. The annual range of temperature is over 16°C. The diurnal range of temperature is also between 14°C to 17°C throughout the year (Fig.2.5). Under the rarefied and least polluted atmosphere, the heat in the day is scorchingly intense and there remains a large difference in the sun and shade temperatures. It is said that if a person sleeps with lower half part of his body in the sun and the upper half part in the shade, then the lower part will get sun burns and the upper part in the shade will have the pneumonia. Even sitting for five minutes in the sun in the months of July and August leads to sun burns at the exposed parts of the body.

FIGURE 2.5

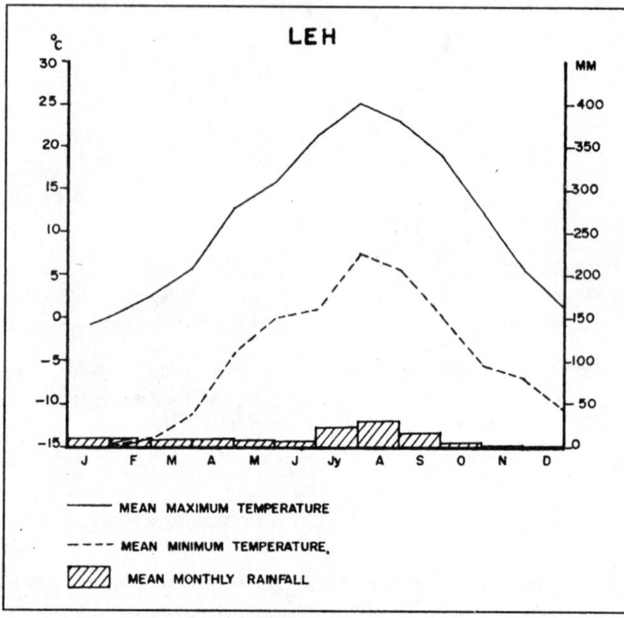

It is not only the annual range of temperature which is significantly high throughout the year, the diurnal range of temperature is also strikingly high in Ladakh in all the twelve months. In the summer months the diurnal range of temperature is of the order of 20°C to 25°C. In winter months the diurnal range of temperature varies between 10° to 20°C. The temperatures recorded for a few days in August 1986 by the author at the Gulmatango pastures of the Suru Valley (Kargil District) have been tabulated and presented in Table 2.4.

The temperature regime in the last week of August 1986 (Table 2.4) reveals a marked diurnal range of temperature (Fig. 2:6).[3] It may be noted from this Table that from 7 a.m. to 3 p.m. there is a steady increase in temperature. This is the period of intense insolation and scorching heat. The actual temperature being 5°C at 7 a.m. which shoots upto 31.25°C at 1 p.m. and declines to 25.5°C at 3 p.m. As the evening approaches, the temperature slumps to about 12°C at 5 p.m. Around midnight it touches the freezing point and reads -1.5°C at 3 a.m. and -0.62 at 5 a.m. The diurnal range of temperature at Gulmatango in the month of August is about 30°C. This is one of the most extreme diurnal ranges of temperatures not only in the Ladakh Division but all over the Indian subcontinent. The high diurnal range of temperature leads to frosting, water freezing in ponds and shallow lakes and accelerates the process of mechanical weathering in the region. It is interest-

FIGURE 2.6

(Suru Valley, District Kargil)
MEAN DAILY TEMPERATURE
August 27-29, 1986

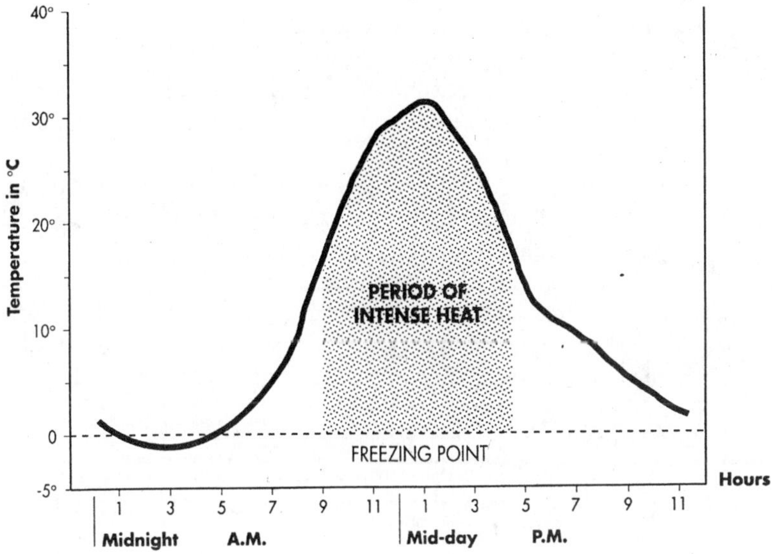

ing to note that the granite which is considered as one of the most hard igneous rocks is weathered more quickly in Ladakh than in any other part of the country.

Table 2.4

Gulmatango (Suru Valley, Kargil District), Mean Daily Temperatures, August 1986 (°C)

Date/Time	*1 a.m.*	*3 a.m.*	*5 a.m.*	*7 a.m.*	*1 p.m.*	*3 p.m.*	*5 p.m.*
27.8.1986	-1	-3	-1	5	32	27	10
28.8. 1986	-1	0.0	-1	6	30	25	12
29.8. 1986	0.5	-2	0.0	4	31	24	14
30.8. 1986	0.0	-1	-0.5	5	32	26	11
Average	0.38	-1.5	-0.62	5	31.25	25.5	11.75

Source: Data gathered by the author during the course of fieldwork at Random and Gulmatango in August, 1986.

In Ladakh, the relative humidity remains low throughout the year. In the driest month (June) the mean relative humidity reads around 30 per cent, and in the dampest months (January and February) only 60 per cent. The amount of rainfall and snow are low. The average annual rainfall is only 3.15 cm. Rain is most frequent in July and August, being 50 and 60 mm respectively. October to December are the driest months, while one or two mild showers are recorded in January and February. In the absence of rainfall, agriculture is almost entirely dependent on irrigation.

The winds are generally light, and depend on the local direction leading to the Khardong Pass, the most common directions are between south and west in the daytime in summer, and from north-east in the night especially in the later months of the year[4].

The rate of evaporation in general is higher to the rate of precipitation in Ladakh. Snowing is quite frequent at high altitudes in winters. The ice-crystals, under the extreme low temperature conditions acquire a highly crystallized shape. The degree of crystallization of snow may be appreciated from the fact that the snow that falls in winter in and around Leh, Kargil and Suru Valley becomes so powdery that it is swept from the flat roofs of houses with brooms. Blowing snow in the months of January and February reduces visibility to a few metres. In the Valley of Nubra, Suru, etc., fog occurs in the morning hours of winters. Summer skies are generally clear. The temperature extremes are the most common cause of weather related illness. In the day time of summer season, one feels dizziness and lassitude. In winters a common direct effect of low temperatures is frostbite. Low temperatures also lead arthritis, swollen sinuses and stiff-joints. Moreover, sudden lowering of temperature puts a severe strain on persons cardiac disorders. The low temperatures, low relative humidity and dry weather are the major constraints in the cultivation of crops in the winter season. In the irrigated fields however, good crops of cereals, fodders, vegetables and flowers are raised. There are numerous orchards of apricot and dry temperate fruits in the valleys of Ladakh.

Notes

1. H.J. Critchfield, 1987, *General Climatology*, 4th ed. Prentice Hall, New Delhi.
2. Husain, M., 1986, *Geography of Jammu and Kashmir,* Rajesh Pub. New Delhi.
3. Husain, M., 1990, 'Suru Vally of Ladakh', *Journal of Tourism*, Lucknow, pp. 35-45.
4. Lawrence, W.R., 1996, *The Valley of Kashmir*, Kishmir Kitabghar, Jammu.

3

■ ■ ■

Natural Vegetation

The term natural vegetation is used loosely to describe any plant life that is not organized or influenced by mankind[1]. The natural vegetation of any region or place is closely influenced by: (i) climate, (ii) physiography, (iii) edephic conditions, (iv) biotic setting, and (v) human interaction with nature. Climate ultimately affects all forms of life. Climate acts in conjunction with other factors to limits of plant growth. Its role is direct in its effects on plants and indirect through its influence on edaphic and biotic factors.

The direct effects of climate on plants are exerted by elements of the water and heat budget, precipitation, humidity, temperature, sunlight and wind. Variation in one can change the others in producing different rates of evapo-transpiration and photosynthesis. The precipitation is also an important determinant of vegetation growth. The deficient moisture limits plant growth, so excess amount restricts certain plants by limiting aeration and oxygen supply in the soil. Excessive soil moisture tends to develop unfavourable soil characteristics and to increase disease damage.

Out of all the climatic factors, the influence of temperature on vegetation is most important. Plants can grow only within certain temperature limits, although the limits are not the same for all plants. For example, certain algae live in hot weather season at more than 90°C and desert lichens withstand 100°C, while arctic mosses and lichens survive -70°C[2]. For each species and each variety, there is a minimum temperature below which growth is not possible, an optimum at which growth is best, and a maximum beyond which growth stops. Most plants, however, cease growth when the soil temperature drops below about 5°C. It is because of this factor that there is hardly any vegetation at 5,550 m (about 18,000 feet) above the sea level in the state of Jammu and Kashmir. The natural vegetation of the state has great altitudinal variations and latitudinal zonation. The spatial distribution of natural vegetation of the state has been briefly presented in the present chapter.

Natural vegetation, often called as "green gold" is an important source

FIGURE 3.1

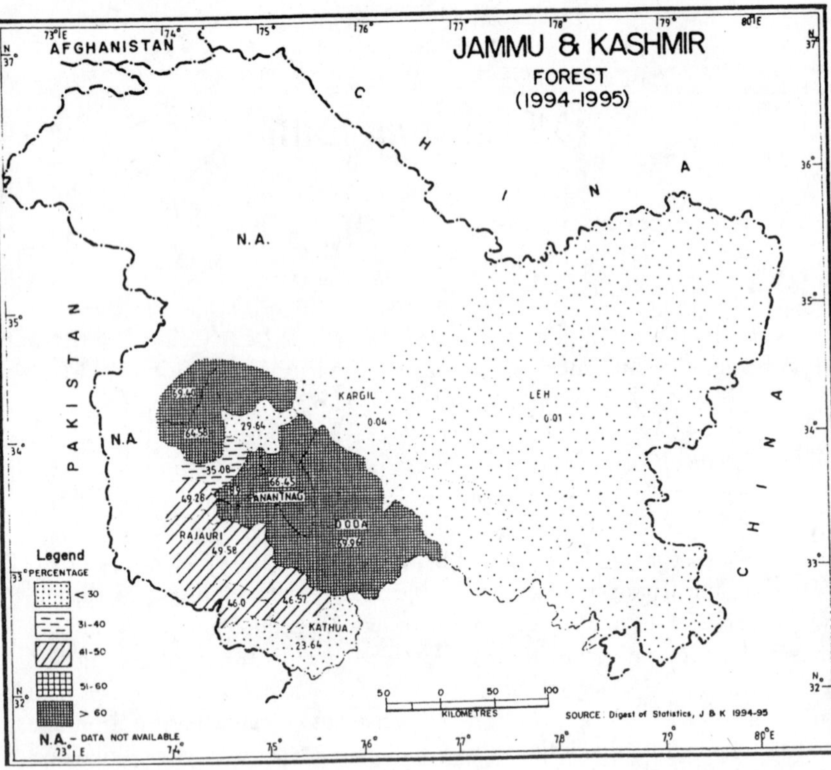

of revenue in the state. The importance of forest in the income of the state may be appreciated from the fact that in 1995-96 it contributed about Rs. 750 thousand millions (75 crores).

The state of Jammu and Kashmir is well endowed in forest resource, still about 33 per cent of the total area of the state is under forest (Fig.3.1). The forests have great diversity in species and varieties, ranging from the lush-green *Margs* (alpine pastures) to evergreen conifers on the gentle slopes of the middle and the Greater Himalayas, and from scrub-jungles of the foothills to the deciduous forests of the Siwaliks and the Pir Panjal Range. The vegetation of the state has however, undergone drastic change. Large tracts of soft-wood (coniferous) and deciduous forests have been removed by continuous felling of trees. Consequently, about 50 per cent of the total forests are degraded and have sparse density of trees. According to the data of 1994-95 the total area under various categories of forests was 20,182 sq km out of which 19,236 sq km was coniferous (soft-wood) and the remaining 946 sq km under deciduous and thorny bushes and scrubs.[3]

The regional distribution of forests in the various administrative divisions (Jammu, Kashmir and Ladakh) has been given in Table 3.1.

Table 3.1

Jammu and Kashmir State: Area under Forest, 1994-95

Division	*Area in sq km*	*Percentage*
Jammu	8124.00	40.25
Kashmir	12039.00	59.70
Ladakh	17.00	0.05
Total	20182.00	100.00

Source: Digest of Statistics 1994-95 Jammu and Kashmir, Directorate of Economics and Statistics, Srinagar.

An examination of Table 3.1 shows that about 60 per cent of the total forest area is in the Kashmir Division and the remaining over 40 per cent is in the Jammu Division. In the Ladakh Division, only about 17 sq km is under forest which may be considered as insignificant. Moreover, the natural vegetation of Ladakh is mainly of xerophytic (vegetation of dry climate) while that of Jammu and Kashmir divisions is mainly either coniferous or deciduous type.

As discussed in the preceding paragraphs, the area stretch, the species and density of forests are directly influenced by lithology, rock-structure, altitude, aspect of slope, insolation and precipitation. The influence of these physical factors is quite pronounced in the state of Jammu and Kashmir. Consequently, there is great diversity in the natural vegetation as over 4,000 species belonging to 1500 genera are found in the state. Some of the important forests of Jammu and Kashmir state consist of deodar willow, juniper, blue-pine, spruce, fir, yew, alder, elm, ash, sorrel, poplar, maple, birch, hazel, and mulberry, silver-fir, cedar, beech, *shisham, mohowa, Jammun,* bamboo, reed, numerous bushes, scrubs and grasses. The district-wise distribution of forest area has been given in Table 3.2.

It may be observed from the Fig. 3.1, (Table 3.2) that the Doda district of Jammu Division has the largest area under forest being 5,848 sq km or about 29 per cent of the total forest area of the state of the Baramulla and Anantnag districts rank second and third, occupying about 13 and 11 per cent of the forest area respectively. The district of Kargil (0.02%) and Leh (0.05%) have the least area under forest in the state. The districts of Udhampur (1.63%), Kupwara (8.18%) and Rajauri (6.46%) are the other areas in which forest cover is reasonably significant while in the remaining districts little or very little area is under forest. Unfortunately, most of the forests of the state are degraded and less effectively managed. According to the National Forest Policy at least one-third of the area of each district should be under forest. Judging from this standard, only Doda, Baramulla and Anantnag districts have around one-third of their total areas under forest.

Table 3.2

Jammu and Kashmir State: District-wise Forest Area, 1994-95

District	*Area under Forest (sq km)*	*Percentage*
Anantnag	2197.00	10.90
Pulwama	729.00	3.61
Srinagar	380.0	1.88
Budgam	481.00	2.38
Baramulla	26677.00	13.26
Kupwara	1651.00	8.18
Leh	11.00	0.05
Kargil	6.00	0.02
Jammu	1339.00	6.63
Udhampur	1945.00	9.63
Kathua	789.00	3.92
Doda	5848.00	28.98
Rajauri	1304.00	6.46
Poonch	825.00	4.10
Total	20182.00	100.00

Source: Computed from Digest of Statistics, Jammu and Kashmir, 1994-95.

Being situated at higher latitudes and characterised by undulating and mountainous topography, most of the forests of the state of Jammu and Kashmir belong to the coniferous category. The specie-wise forest area has been given in Table 3.3.

An analysis of Table 3.3 vividly shows that about 37 per cent of the total forest area is under conifers, comprising of *deodar* (oak), *chir* (pine), *kail* (blue-pine) and fir, and about 50 per cent is under miscellaneous trees of deciduous and conifers. The deciduous trees of teak, *sal*, *jammun*, *mahuwa*, *shisham*, etc., constitute about 4.27 per cent of the forest area, the deciduous forests are mainly confine to the Siwaliks and the Lower Himalayas. About 9 per cent of the forest area is under the category of wildlife reserves.

Table 3.3

Jammu and Kashmir State: Specie-wise Forest Area, 1994-95

	Type of Forest	*Area in sq km*	*Percentage*
1.	Coniferous		
	a. Deodar	1122.48	5.07
	b. Chir	1773.03	8.00
	c. Kail	1876.46	8.47
	d. Fir	3355.43	15.15
	e. Others	11108.40	50.18
	Total coniferous	19235.80	
2.	Deciduous	946.20	4.27
3.	Wildlife reserves	1962.90	8.86
	Total	22144.90	100.00

Source: Computed from the Digest of Statistics 1994-95 Govt. of Jammu and Kashmir, p.81.

The Lower and the Greater Himalayas are quite rich in timber forest. The most valuable timber, that of *deodar* (oak) is mainly found in Baramulla, Anantnag, Doda and Udhampur districts. The system of farming-out timber to contractors has mainly resulted in the destruction of many fine forests in the vicinity of rivers, and the tract which the contractors have spared have suffered at the hands of Gujjars and Bakarwals, who with their small axes cut down trees partly for the sake of fodder and partly from a kind of natural instinct which impels them to make forest clearings for the sake of grass. Dry timber is harder to cut up than green grass, and the contractor, instead of using trees which have felled, prefers to cut down fresh ones[4].

The best of all timber is obtained from deodar. It is much in demand for house, boats, and bridges as it is impervious to water. The old shrines, mosques, temples in Kashmir and the hilly areas are made of deodar. The blue-pine is much valued as a timber for house-building. It also yields an excellent charcoal. The spruce timber is also used for house-building. It is however, not durable when exposed to water and moisture. The silver fir, yew, and elm are also used for house-construction, while alder, used for furniture and walnut is in great demand for furniture, gunstocks, and spinning-wheels. The maple, elm, poplar are used for agricultural implements while willow is used for sports goods and basket making. The *chinar* (boin) is a royal tree in Kashmir and much appreciated as a shade tree. The mulberry wood is mainly used for the doors and agricultural implements like plough[5].

As stated at the outset, the climatic factors exert great control over natural vegetation. In other words, there is a close relationship between the average annual temperature, average annual precipitation and the type of natural vegetation. In fact, the climatic belts overlap the vegetation belts in the world. In an essentially mountainous state like Jammu and Kashmir, there are micro-level variations in climatic factors. Consequently, the natural vegetation differs dramatically form place to place. Moreover, the percentages of leaves, stems, and roots, etc., also differ from forest to forest. In order to understand the spatial distribution of different types of natural vegetation, the forests of Jammu and Kashmir state may be classified under the following categories.

1. Subtropical Forests

The subtropical rain forests are confined to the Siwaliks and lower slopes of the Middle Himalayas. Owing to the seasonality of rainfall and edaphic factors there is great diversity of trees in these forest. These factors have a thick undergrowth of bushes and scrubs. The dominant species of subtropical forests are teak, sal, *Shisham*, *Pipal* (ficus religiosa), *Tun*, silver-pine, *Mohowa*, *Khair* (Acacia-catehu), thorny bushes, evergreen shrubs, climbers, reed and tall grasses locally known as *Khar*. Most of these species are broad-leaved deciduous type which drop their leaves in the months of January, February and March, just before the advent of hot weather season. In the biomass of subtropical forests the proportion of leaves is 8 per cent and the stems 74 per cent, while the remaining 18 per cent is roots[6]. These forests are mainly utilised for fuelwood, timber, house-construction, agricultural implements and miscellaneous purposes. Valuable products like resins, gum, *Katha*, and medicinal herbs are also obtained from the subtropical forests of the state. These forests are being destroyed and depleted at a faster pace. The pressure of human population and ignorance of people are quickly damaging the ecosystems of these forests.

2. Temperate Forests

Moving northwards as the latitude and altitude increase the place of the subtropical forest is taken by temperate vegetation. The slopes of Pir Panjal, Greater Himalayas, Zanskar and Karakoram between 1,500 m to 3,000 m are dominated by temperate forests. The dominant species in these forest are *deodar* (cedrus Deodara), pine (pinus), silver-fir, spruce, fir, elm, alder, cedar, ash, sorrel, birch, paper-birch, and hazel. The northern slopes of Pir Panjal have the dominance of *deodar*, blue-pine with undergrowth of deciduous shrubs, while the Jhelum Forest Division (stretching over Gulmarg and Lolab Valley) is dominated by cedar, fir and spruce species of conifers. In the valley floor of Kashmir, poplar, *chinar*, maple and *vir* (willow) are the main species of vegetation which are deciduous in character. Where soil conditions permit, mixture of broad-leafed deciduous trees, such as maples and oaks, grow together.

The temperate forests are utilized for timber, fuelwood, charcoal, and house construction. These forests produce good quality of timber and paper-pulp. The most valuable timber forests of deodar have been depleted and are now confined to the north-western corner of Kashmir Division. The system of farming-out timber to contractors has resulted in the destruction of many good forests. The tracts which the contractors have spared, suffered at the hands of Gujjars who with small axe cut down trees partly for the sake of fuelwood and fodder and partly from a kind of natural instinct which impels them to make forest clearing for the sake of grass. Moreover, the inadequacy of hydel and thermal power, especially during the severe winters, compels the people to fell the trees and their twigs for fuelwood to be burnt in the winters to keep the house warm and cozy. The accelerated rate of felling of trees is leading to the shrinkage of area of deodar, spruce, pine and fir. The overexploitation of forests and the use of soft and costly wood for fuel are doing great harm to the state revenue on the one hand and to aesthetic beauty and ecosystems on the other. In fact, many of the ecosystems have lost their resilience characteristics.[7]

3. Alpine Pastures

In the state of Jammu and Kashmir the vegetation, especially grasses belonging to the higher regions of mountain system is known as *Margs* or alpine pastures. The alpine pastures lie between 3600 m to 4000 m above the sea level. Being the zone of high altitudes, the climate in alpine pastures is extremely cold over greater parts of the year. It is only in the summer months (May to September) when ice melts at high altitudes and lush green grasses grow. The low temperatures at high altitudes of the Himalayan belt of the state support some dwarf varieties of birch and junipers making a shrubby appearance. The stunted conifer trees merge into extensive alpine pastures. In these pastures, some shrubs form isolated thickets during summer months (June-August). The lush green and nutritious grasses of alpine pastures are utilized and grazed by the Gujjars and Bakarwals (goat herders) who practice transhumance. These herders ascend in the alpine pasture with their goats and sheep. They stay in *Margs* pastures upto the middle of September, depending on the prevailing temperature and precipitation conditions. In the autumn season (September-October) they descend and migrate with their flock to the zones of lower altitudes—kani areas of the Jammu Division to pass their winters.

Natural Vegetation of Ladakh

In Ladakh, the average annual rainfall is less than 20 cm. Ladakh is a high altitude desert. Lack of moisture is the essential factor shaping the vegetation of Ladakh. The absence of rainfall over greater part of the year has resulted into xerophytic vegetation. The plants of Ladakh have evolved many special characteristics for conserving water. In plants and trees these include usually thick, waterproof outer layers or cuticles, modification of breathing pores, reduction in leaf area, and specialized hairs or outgrowths that reflect light. Many plants of Ladakh, like that of deserts, are heavily armed with spines to repel the attacks of moisture seeking animals. Thus, they have biochemical defence against herbivores. In certain soil conditions, Ladakh plants have extremely long tap roots to reach deep underground water supplies. At the occurrence of rain, the soil tracts of Ladakh appear to turn miraculously green almost overnight, and fast growing, short lifecycle plants often create spectacular floral display.

The natural vegetation in Ladakh is absent over the greater part except for more humid region of Nubra and other side valleys of the Indus river. Hippophae-scrub is the major woody component which is used as fuel and material of roofing of houses.

The above description shows that the state of Jammu and Kashmir is relatively poor in natural vegetation as compared to the other Himalayan states of India. The growing pressure of population, the contractual system of felling trees, over-interaction of man and poor management are responsible for the depletion of forests. During the last 50 years, a substantial proportion of forests have been cleared and brought under agriculture and pastures. Whatever, is left of the former vegetal cover, visa-a-vis and kinds of stresses and strains in the state still has a great important in regard to soil conservation, besides providing timber, fuelwood, fodder, medicinal herbs and water and aesthetic beauty conservation. Looking at the benefits accrued from forests, their conservation and judicious utilization deserve priority. To achieve these objectives a sound pragmatic strategy needs to be evolved in consultation with the researchers, planners, administrators and rural masses. Some of the steps which may help in reducing the rate of depletion of forests have been given as under:

1. The government should implement the National Forest Policy strictly and the wrong doers be brought to the book. In order to reduce the pressure on forest, alternate sources of power (hydel and thermal) be adequately developed to meet the growing demand of fuel for cooking and heating in the rural and urban masses.
2. The forests may be conserved by preventing the wastes in lumbering. Still at many places, much of the waste material is burnt which can be

used for fuel and some in making small articles like toys, decoration pieces and handles for tools. More attention is to be paid to the treatment of lumber with chemicals to prevent its decay and destruction by insects.

3. Each year forest fires sweep over large areas. This great waste can almost wholly be prevented if adequate attention is paid to safeguard the forests from fire. In dry season, especially in summers, hunters and idle visitors should be forbidden to enter the forests. The burning of forest waste should be done after a heavy rain or snowfall when the danger of conflagration of fire is minimum.
4. Modern technology for harvesting forest should be employed.
5. The community land should be brought under social forestry to meet the fuel and fodder demands of the local people.
6. Reforestation should be attempted on barren lands and undulating slopes
7. A comprehensive forest management and development plan has to be adopted to increase the productively of forests, to link up forest development with forest based industries and to develop forests as supports to economy.

Wild Life

The undomesticated animal life of a region is known as wild life. The state of Jammu and Kashmir has great diversity in the physiography, climate, habitat and natural vegetation. This diversity in the geo-ecological conditions provides the required habitats for different animal kingdoms. The wild life in Jammu and Kashmir, consequently, differs from south to north and from lower altitudes to higher altitudes. In general, leopard, tiger, bears, wolf, foxes, deer, hangar, stag, wild cats, moose, rabbits, squirrels and a variety of birds are the main fauna in the state. At high altitudes are found warm-blooded, fir animals and musk-deers.

Musk-deer (*Kasturi-mrig*), monkeys, langoors, wild pig (*Bad*) have their habitats between 2,000 and 3,500 m above the sea level. They, however, shift their habitat seasonally. In winters, they descend at lower heights and in summers ascend to higher altitudes. The Central Himalayas, Zanskar Range, and higher reaches of Pir Panjal and Dhauladhar are the abodes of these animals. The pod (*Nafa*) of Musk-deer contains a soft, brownish matter which is the much coveted *Musk*; about 20 grams in weight and the price rules high. Consequently, the people risk a good deal to secure it[8].

Leopard, *Cheetah* (tiger), wolves and foxes are found in the forest of Pir Panjal, Liddar (Anantnag Dist), Doda, Udhampur, Riasi, Uri, Poonch, Karnah, Kupwara, Daksum, Sindh and Naranag catchments. Black-bear is very common in the valleys of Lolab and Liddar, Doda, Rajauri and Uri forest. Stag (*Hangal* or *Barasingha*) are also found in these localities.

Snow-partridges, Himalayan-cock, chikor (*Kak*), cuckoo, larks, bulbuls, quil, storks, cranes, ducks, geese, hawks, parrots, pigeons, dove, ducks, *teel*, vultures, crow, eagle and owls are the main birds of the state. A wide variety of reptiles, snakes, fish, and insects are also found in the various parts of Jammu and Kashmir state.

Notes

1. J. Whittow, *Dictionary of Physical Geography*, Penguin Books,
2. Critchfield, H.J., 1987, *General Climatology*, Prentice Hall of India, New Delhi.
3. *Digest of Statistics*, 1993-94, Govt. of Jammu and Kashmir, Srinagar pp. 80-81.
4. Lawrance, W.R., 1996, *The Valley of Kashmir*, Kashmir Kitabghar, Jammu.
5. Ibid. pp 80-82.
6. Ehrlich, P.R. et.al. 1977, *Population, Resources and Environment*, Freeman & Co. San Francisco, p. 146.
7. Husain, M., 1986, *Geography of Jammu and Kashmir*, Rajesh Pub. New Delhi, p. 38.
8. Lawrance, W.R. op cit. pp. 106-140.

4

■ ■ ■

Soils

The naturally occurring thin layer of unconsolidated material on the earth's surface that has been influenced by the parent material, climate, relief, and physical, chemical, and biological agents is known as soil[1]. Soils exhibit differences in their physical and chemical characteristic (soil structure, soil temperature, soil texture, soil humus, soil water), as well as in their capability for growing crops soil may be classified into: (i) zonal, in broad latitudinal zones or belts, (ii) inter-zonal, resulting from special parent rocks, (i.e., limestone, sandstone), (iii) azonal or immature soil, new material on which soil forming processes have not had sufficient time to work.[2] In brief, differences in soils arise from the mineral composition of the parent materials and from differing climatic conditions, which together influence the organic and inorganic processes of soil development.

A comprehensive soil map of Jammu and Kashmir state, based on scientific data, has not been prepared so far. The available soil maps of the state give a parochial picture of the soils. The land revenue records and settlement reports have classified soils on the basis of texture, structure and colour, while the rural folk and cultivators recognize their soils by different nomenclatures which differ from slope to slope and in many cases from village to village.

Under the existing geo-climatic conditions, a wide range of soils, both of residual and alluvial origin are found in the state. As discussed in the opening chapter of this book, the state of Jammu and Kashmir is essentially hilly and mountainous. The outer plain of Jammu with alluvial soils was deposited by running water and the fluvio-glacial action. The hilly and mountainous areas are generally covered by the residual soils, while the upper reaches of Chenab and the Jhelum and their tributaries are covered with alluvial and morainic soils. On the basis of rock strata and pedogenic characters, the following major categories of soils may be identified in the state.

Hilly and Mountainous Soils

The hilly and mountainous soils are found in the entire state excepting the levelled plain of Jammu adjacent to Punjab and the valley floor of Kashmir. The undulating topography and steep slopes effect the run-off and drainage system. Other things being equal, the run-off is large on steep slopes. As a rule, more the water run-off, lesser is its absorption in the soil on steep slopes. The run-off also washes away more of the weathered rocks on steeper slopes. The depth of soil and soil profiles on steeper slopes are consequently shallower than that on gentle slopes. Many soils in the mountainous areas are shallow, immature and highly susceptible to soil erosion. These soils are generally, acidic in character, deficient in potash, phosphoric and lime and therefore, need regular manuring and fertilization for good yields. The humus content in these soils varies from slope to slope and altitude to altitude. Depending on the availability of sunshine and rains, these soils are generally devoted to the cultivation of maize, pulses, orchards (almonds, apples, peach, pears), oilseeds, barley, wheat, oats and fodder. The higher altitudes, between 2500 m to 4000 m are reserved as alpine pastures. These pastures develop nutritious grasses during the summer season which are grazed and utilized by the Gujjars and Bakarwals for their flocks of goats, sheep and horses. Scarcity of water, leaching, erosion and avalanches are the main problems of these soils. These problems are getting accentuated owing to the indiscriminate felling of trees and depletion of ecosystems.

Alluvial Soils

The alluvial soils are deposited by the action of rivers. They are found in the river channels, floodplains, estuaries, lakes and fans at the foot of mountains. The alluvial soil includes all consolidated fragmented material from the coarsest gravels and sands down to the finest clay and silt-sized particles. In other words, sand, silt and mud brought down by rivers in floods and deposited on the temporarily submerged land are known as the alluvial soils. These are the most productive soils of the state, found mainly in the Jammu-plain, the valley floor of Kashmir and at narrow river terraces along the tributaries of the Chenab, Jhelum and their tributaries. The alluvial soils may be classified into: (i) the old alluviums, and (ii) the new alluviums. The old alluviums lie above the banks of rivers and are generally free from floods, while the newer alluviums are frequently inundated as they lie in the flood plains of the Jhelum, Chenab and their tributaries. New alluviums (*Khadar*) soils are relatively coarse in texture and contain more sand than that of old alluviums (*Bhangar*) soils. The alluvial soils, wherever, irrigated grow two to three crops in a year. In the Valley of Kashmir, these are largely devoted to paddy, maize and orchards, while in the Jammu plain they are

FIGURE 4.1

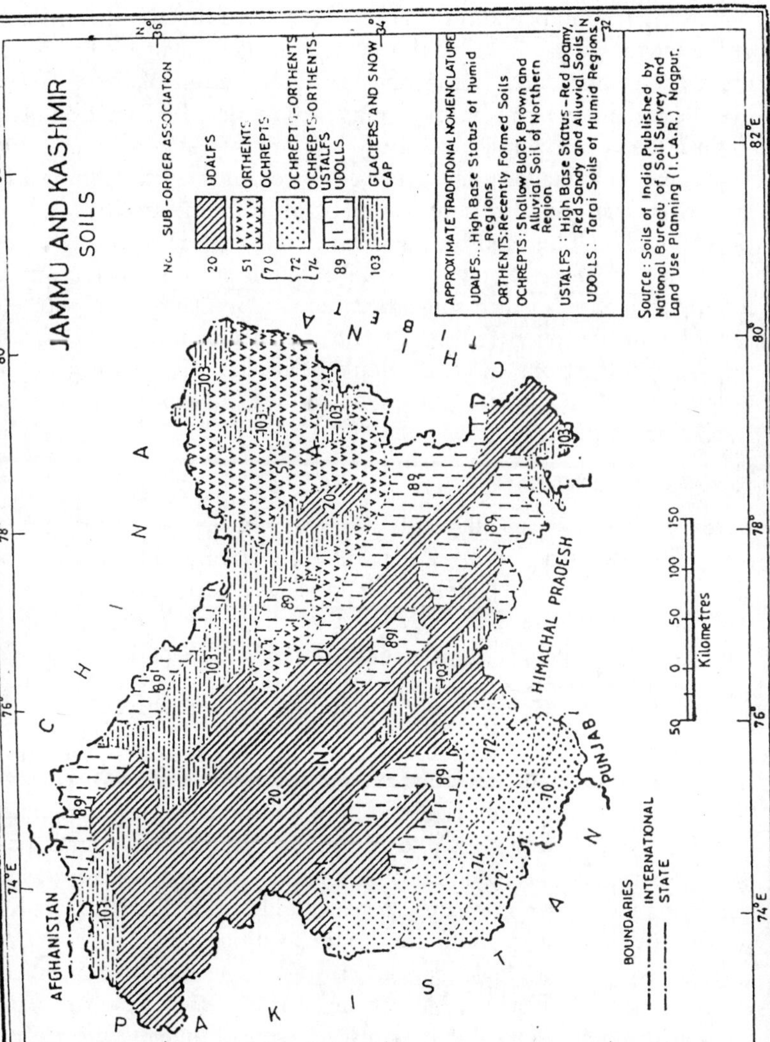
JAMMU AND KASHMIR
SOILS
NO. SUB-ORDER ASSOCIATION
20 UDALFS
51 ORTHENTS OCHREPTS
70 OCHREPTS-ORTHENTS
72 74 OCHREPTS-ORTHENTS USTALFS
89 UDOLLS
103 GLACIERS AND SNOW CAP
APPROXIMATE TRADITIONAL NOMENCLATURE
UDALFS...High Base Status of Humid Regions
ORTHENTS: Recently Formed Soils
OCHREPTS: Shallow Black Brown and Alluvial Soil of Northern Region
USTALFS : High Base Status - Red Loamy, Red Sandy and Alluvial Soils
UDOLLS : Tarai Soils of Humid Regions
Source: Soils of India Published by National Bureau of Soil Survey and Land Use Planning (I.C.A.R.) Nagpur.
BOUNDARIES
INTERNATIONAL
STATE
Kilometres
AFGHANISTAN
CHINA
PAKISTAN
INDIA
TIBET
HIMACHAL PRADESH
PUNJAB

utilized for the cultivation of wheat, gram, pulse, paddy, mustard, barley, sugarcane, oilseeds, potato, barseen, vegetation and fodder crops. Being highly productive, these soils are giving good returns of the High Yielding Varieties of wheat and rice in the areas of controlled and assured irrigation.

In the Jammu plain, in the areas of canal and tubewell irrigation, there occur intrazonal saline soils. The saline and alkaline affected soils become useless and unproductive from the agricultural point of view. These soils may be reclaimed by mechanical and chemical devices. Irrigation is needed to flush the salts down through the soil to prevent them from accumulating.

Saline and alkaline soils can also be reclaimed by the removal of top layer of the soil and by the application of farmyard manure, tank silt, molasses, gypsum, sulphur and by the cultivation of green manuring crops like *dhaincha*, *sani*, *barseem* and legumes. Plantation of trees also helps in reducing the formation of such soils. Rice, lucerne, and barley can withstand to some extent the adverse effect of saline and alkaline formations. Moreover, seepage of water from the canals should be checked. The drainage of the waterlogged areas needs improvement to control the expansion of saline and alkaline formations.

Karewa Soil (Wudur)

Karewas are fresh-water (fluviatile and lacustrine) deposits found as low flat mounds or elevated plateaus in the Valley of Kashmir and the Kishtwar and Bhadawah tracts of the Jammu Division. The important Karewas are found in Kulgam, Shopiyan, Badgam, Qazigund, Tanmarg, Gulmarg, Baramulla, Laithpora, Chandhara, Pampore, Bijbehara, Awantipora, Islamabad (Anantnag), Mattan, Tral and Ganderbal.

The Karewa soils are composed of fine, silty clays with sand bouldry gravel, the coarse detritus being as a rule, restricted to the peripheral parts of the valley, while the finer variety prevails towards the central parts.[3]

The Karewa soils of Kashmir have enormous agricultural potential. Commercial and cash crops like saffron, almond, apples, walnut, peaches, pears, cherry, plum, etc., with orchards and saffron beds. Moreover, some leguminous and fodder crops are also grown in Karewa. The pampore Karewa is famous all over the world for saffron cultivation.

Soil erosion and depleting soil fertility are the major problems of the karewa soils. It has been reported by the farmers of the Chandhara and Dusu villages (Pampore-Karewa) that with the passage of time the karewa soils are losing their resilience characteristics. As a matter of fact, the per unit production of saffron and almond has gone down substantially during the last three decades. The soil conservation practices need to be adopted to maintain the health of the karewa soils, making them economically more productive and ecologically more sustainable.

The Kashmir farmers on the basis of their empirical experience recognize four classes of soils. These classes are known as: (i) Gruti, (ii) Behil, (iii) Sekil, and (iv) Dazanlad[4].

Gruti (Clayey Soil)

Gruti soils contain a large proportion of clay. Texturally, it resembles to the clayey loam. Its water retaining capacity is high. In years of scanty rainfall, it is considered to be the safest for the cultivation of rice. Contrary to this, if rains are heavy, the gruti soil gets compacted and achieves the shape of hard cakes, the ploughing of which becomes difficult and pulverization of soil is an arduous task. In the years of scanty rainfall, these soils give poor yields. The gruti soils are found in the low-lying areas of the Kashmir Valley.

Behil (Loamy Soil)

Behil is a rich loam of great natural fertility. The humus content is high which enriches the soil fertility. Consequently, it does not require heavy manuring. Moreover, there is always a danger that by overmanuring the soil will be too strong, in which the rice crop will show more vegetative growth and will be more susceptible to lodging. It is ideally suited for paddy cultivation.

Sekil (Sandy Loam)

Sekil is a light loam with sandy subsoil. In the sekil soil field if artificial irrigation is available, good crops of rice are harvested in the summer season. Sekil soil is generally confined to the lower edges of karewas in the Valley of Kashmir.

Dazanlad (Sandy Silt)

Dazanlad soil is chiefly found in the low-lying ground near the swamps, but it sometimes occurs in the higher villages also. The soil has an admixture of sand and clay and becomes more warm in the summer season. A peculiar characteristic of dazanlad is that the irrigation water when stands in the fields turns red in colour. If controlled irrigation is provided, high yielding varieties of rice can be grown successfully in dazanlad soils.

Nambal (Peaty Soils)

Near the banks of the Jhelum river and in the vicinity of the Wular, Manasbal and Anchar lakes is found the rich peaty soil, locally known as Nambal. In the years of normal rainfall and moderate snowfall, nambal soils give good yields of rape-seed, mustard, maize, oats, pulses and fodder.

Tand (Mountainous Soils)

The land on the slopes of mountains, reclaimed from the forests is called *Tand* soil. After reclamation the tand gives good yields of maize, pulses and fodder for two or three years, but under the impact of accelerated soil erosion the land loses its natural strength. Consequently, the productivity declines and after a period of about six to ten years the land acquires the shape of a pasture and culturable waste.

Zabelzamin (Alkaline Soils)

Patches of irrigated land if excessively irrigated lose their fertility and develop alkaline formations. Such adversely affected patches of saline and alkaline formations are known as *zabelzamin*. These soils are unproductive from the agricultural point of view unless especially treated with gypsum, water and manures.

There are numerous other types of soils recognized by the Kashmiri farmers, such soils are *Kharzamin*, *Tresh*, *Limb*, *Ront*, *Shath* and *Tats*.

Soils of Ladakh

Excepting the soils deposited by the rivers and the glaciers, the soils of Ladakh are unique and have largely mechanical origin. In Ladakh, the highly weathered rocks, under the mechanical impact of insulation and frosting result into fine-grained, light-coloured soils which are deficient in humus contents. These soils are however, rich in potassium, nitrate, and phosphate and give excellent agricultural returns in the tracts where irrigation water is available.[5]

On the river terraces of Indus and its tributaries, the soils are of alluvial nature which are being utilized for the cultivation of vegetables, grim (a type of barley), wheat, small millets, pulses, fodder, apricot and orchards.

Soil Conservation

The depletion of soil owing to physical and cultural factors is a universal phenomenon. The rate of soil erosion and soil depletion in a given region is largely influenced by: (i) rainfall erosivity, (ii) volume of run-off, (iii) wind-strength, (iv) relief, (v) angle of slope, (vi) slope length, (vii) length of wind fetch (viii) shelter belts, (ix) the mode of soil utilization, (x) intensification of agriculture, and (xi) management of land for agriculture and allied economic activities. The rate of soil erosion also depends on the pressure of population, cropping patterns, cropping intensity, crop rotation, irrigation, manuring and tillage practices.[6]

Soil erosion not only affects the areas from which soil is removed but also influences the environment where it is deposited. Such deposits if take place in lakes and ponds, destroy the aquatic remedial measures to combat soil erosion and to keep the ecosystems in a healthy condition so that the needs of the present generation be fulfilled without compromising the needs of the future generations. Some of the steps which may go a long way in overcoming the problem of soil erosion are given as under:

1. Soil erosion is more serious in the areas where deforestation has taken place at a large scale. Forest ecosystems exercise considerable control over pattern of climate, hydrology, circulation of nutrients, erosion and cleansing functions of air and water, as well as over the status of streams, lakes, and underground water supplies. Forests also present a vast reservoir of genetic diversity of plants, animals and micro-organisms, which play significant roles in maintaining the ecosystems.
2. It is imperative to maintain proper records of land productivity status which should be prepared with the help of geographers and soil scientists.
3. In areas of high irrigation requirement like paddy growing areas, effective soil and water conservation techniques should be adopted with the help of technical experts and agricultural scientists.
4. Sustainable cropping patterns and rotation of crops should be followed by the cultivators. Cultivation of soil exhaustive crops in consecutive years should be discouraged for which extension service programme is to be launched effectively.
5. In the areas of water erosion, the agroforest ecosystem is to be implemented and strengthened.
6. Industrial activities in the nearby areas which are hazardous to the environment should be regulated through law.
7. The culturable waste and marginal lands should be amended properly so that further degradation of such lands may be checked.
8. In the soil erosion areas, adoption of integrated watershed management system need to be implemented.
9. The auto-regenerating of the soil fertility by adding organic matter through micro organism and using crop residues should be facilitated.
10. Pressure of population on arable land should be reduced by implementing effective population policy.

Notes

1. J. Whittow, 1984, *The Dictionary of Physical Geography*, Penguin Books.
2. L. Symons, 1969, *Agricultural Geography*, London.
3. Husain, M. 1996, *Systematic Agricultural Geography*, Rawat Pub. Jaipur.
4. Husain, M. 1986, *Geography of Jammu and Kashmir*, Rajesh Pub. New Delhi.
5. Ibid.
6. Plyurnin, *Reclamation, Soil Science*, New York.

ties and moisture contents.

The Geological Survey of India has carried out explorations at Mohogala and Metka (Poonch District) and arrived at the result that these places, upto a depth of 300 m have about 9 million of coal. The fixed carbon in the coal of these deposits is about 57 per cent, volatile 30 per cent and ash and moisture 10 and 3 per cent respectively.

At present coal is being mined near Kalakote to feed the only thermal power plant of the state at Kalakote. The rated capacity of the plant is 7.5 MW and 35,000 tonnes of coal is being mined annually in its vicinity. The energy generation capacity of the plant may be enhanced substantially if new generators are installed in the Kalakote Plant and new technology is applied for the mining of coal.

Lignite deposits are found mainly in the Valley of Kashmir which occur in the Karewa formations, right from Nichahom upto Lolab. The lignite seams are, however, associated with clays and loams and are interbedded in carbonaceous clays in various localities of Kashmir Division. The major lignite deposits are in are in the vicinity of Nichahom, Chowkibal, Budhasheng, Lanylab, Shaliganga, Raithan and Tangmarg.

According to the Geological Survey of India, Nichahom lignite seams occur from north to south, running over a distance of about 85 km and the estimated reserve of lignite in this mine is about 85 km and covering a width of 16 km upto a depth of 8 m. The estimated reserve of lignite in this mine is about 6 million tonnes. In between Nichahom and Chowkibal a reserve of 8 million tonnes has been estimated.

The total estimated reserve of lignite deposits in the state, according to the Directorate of Geology and Mining is about 85 million tonnes, but owing to thin seams and high ash contents, it is mined only at Nichahom in Kashmir. In fact, the lignite of Kashmir Valley contains only 8 to 25 per cent carbon, 30 per cent ash, 15 per cent moisture and 27 per cent volatile matter.

Peat—a superficial accumulation consisting of vegetation matter which has become decomposed to a certain limited extent in the cold regions is also found in the Valley of Kashmir. Its deposits are found mainly in the swampy grounds on both the sides of the Jhelum river below Srinagar. Peat is cut and dried before being used as fuel. It is applied in field as a manure. The Kashmiris call peat as demb-tsak.

Natural Gas

Natural gas has not been struck so far in any part of the state. In Jammu Division, areas like Ramnagar, Dharmthal, Nauashara and Rajauri are under investigation for natural gas. Occurrence of dry gas has been reported by the Oil and Natural Gas Commission (ONGC) in the Kashmir Division. The

ONGC has proved the existence of *'Dry Gas'* at the Silk Factory Road, Rambagh (Srinagar). Various indications such as sulphurous water, hot springs, gas seepage, etc., have been noticed in several areas of Jammu and Kashmir which prove that in near future oil and natural gas may be found in the state.

Metallic Minerals

There are a few metallic minerals found in the state. Though at present the state is producing very little metallic minerals, there are evidences which prove that gold and copper have been mined in the Ladakh and Kashmir divisions respectively sine the medieval period. The occurrence of metallic minerals in the state is in isolated tracts and in many cases their exploitation is uneconomic. A brief description of some of the metallic minerals found in the state is given in the following paragraphs.

Iron-ore

Iron-ore is obtained from hematite and magnetic mainly found in igneous rocks. The state of Jammu and Kashmir does not possess significant deposits of iron-ore. There are however, some workable iron-ore deposits in the Riasi Tehsil of Udhampur district, in association with the Nummulitic Series, which supported a number of local furnaces for the manufacture of munitions of war during the last two centuries. Still no area of iron-ore has been identified in the Kashmir and Ladakh divisions of the state. The non-availability of iron-ore is one of the main reasons for the industrial backwardness of the state.

Gold

Gold is one of the rare and precious minerals. It is found in hydrothermal veins in the crust or in alluvial grains, placer deposits and conglomerate.

Gold bearing quartz veins have been found in areas of Dras, Suru and Indus Valley in the Kargil and Leh districts of Ladakh Division. The gold found in quartz veins gets exposed under the impact of weathering and may be transported by running waster. In fact, particles of gold have been noted as placer deposits in the course of rivers and streams of Ladakh and the terraces of Indus River. Gold has been traced in conglomerate deposits in the valleys of Dras, Suru and Indus, forming a contiguous deposit, ranging in thickness from one metre to four metres. It is estimated that 0.6 gram of gold can be obtained per cubic metre of conglomerate in Ladakh. Alluvial gold-washing is carried on in the sands and gravels of many of the tributaries of the Indus river.

Copper

Copper is extracted from the ore of chalcopyrite which is found in rocks like serpentine, diorite, chloriteschist and slates. Chalcopyrite a double sulphide of copper and iron is the main copper-ore found at Sumbhar (Anantnag), Lashtial (Baramulla) and Ganderbal. In Ladakh region native type of copper has been reported from areas of Zanskar. Investigations about the occurrence of copper in Doda District have also been started. It is hoped that copper-ore in all the quartz veins of quartzite rocks of the Doda District have appreciable quantities of copper which may be extracted economically.

Lead

Lead does not occur in nature as the native metal, yet it has been known to man since the earliest times because of its easy extraction from the ore. It is a mineral of great economic importance as it is used largely in numerous industries.

The main ore of lead is galena. It is found in hydrothermal veins. Galena is a grey coloured mineral. It has shining metallic lustre, relatively soft and very heavy, often crystallizing in nearly perfect cubes. Galena and other ores of lead are found in association with slate, phyllites graphite-schist and limestone near Ramsu (Jammu Division). About one million tonnes of galena has been estimated in the Ramsu region of the Doda district.

In the Division of Kashmir, galena deposits have been reported in Buniyar (Baramulla). The ore of Buniyar deposits may yield 30 to 70 per cent of lead. Several parts of side-valleys of Kashmir are being investigated and there is a strong possibility of lead deposits in the Panjal Trap formations.

Bauxite

Bauxite is a residual deposit formed under special climatic conditions. It is a mixture of hydrated aluminium oxides, but the name is applied particularly to varieties of the rock laterite rich in hydroxides of aluminium which are the chief ores of the mineral.

Bauxite is a mineral of great economic importance. Being light and resistant, it is used mainly in aircraft industry. In the state of Jammu and Kashmir extensive deposits of bauxite are found in Chhakar (Riasi), near Songarmarg, Salal, Panhasa, Baladanu, Sangarmarg, Sukhwal-gali and Khander in the districts of Rajauri and Udhampur. The Geological Survey of India has estimated about 12 million tonnes of bauxite in these fields and the analysis has revealed that the ore contains 60 to 70 per cent of alumina.

In 1994-95, the production of bauxite was over 2125 tonnes. A calcined bauxite unit set up by the Kashmissa Industries at Jammu, started production in 1983.

Chromium

Chromite is used in the manufacturing of refractory bricks for furnace-linings. Its chief metallurgical use lies in its being the raw material of chromium. An alloy of chromium and iron (ferro-chrome) is used in the making of rustless and stainless steels and armour plates. A large amount of chromium is used in the manufacture of mordants and pigments, because of the red, yellow, and green colours of its salts.

The main ore for chromium is the chromite which is found in ultrabasic igneous rocks and serpentinites. The chromite-ore occurs usually as brown or black crystal. In the state, chromite deposits are found in the cretaceous volcanis of Burzil, Dras, Bumbat, Tashgam, and Kargil Valley of Ladakh. It is mainly used in metallurgy.

Zinc

Zinc is obtained from the zinc sulphide and zinc blend. The mineral is found usually in limestone and veins associated with copper and lead ores. Zinc is found in Buniyar area in association with galena, chalcopyrite and pyrites. Zinc ore is also found in limestones deposits of Udhampur.

Sapphire

Sapphire is a blue transparent or translucent corundum (gem). It is naturally occurring oxide of aluminium and contains as much as 52 per cent of aluminium. In India, Jammu and Kashmir is the only state in which sapphire deposits of good quality are found. The famous sapphire mines of poddar near Sumjam in Kishtwar are located at an elevation of 4,500 m in the upper reaches of Chenab River in the Doda District. In Kishtwar sapphire was discovered in 1882, and since then mining has been going on intermittently. The precious stone is interbedded in the rocks of Salkhala Series and is associated with meta-basic, intruded by a number of pegmatite and quartz veins. The sapphire bearing portions are lenticulars composed of plagioclase, felspars, tourmaline and corundum. The precious stones occur as small prism of 5 mm width and show pyramidal structure.

Transparent crystallized corundum occurs in pegmatite veins cutting actinolite-schist lenticles in Salkhala marble at an altitude of 4,500 m. Associated minerals in the pegmatite are prehnite, tourmaline, beryl, spodumene and lazurite. The Mineral Survey of Kashmir has revealed a large quantity of crystallized transparent corundum. The bulk of the output from the mines is confined to what are called 'rock sapphires' valueless for gems and of use as abrasives, watch jewels, etc.[7]

Ruby

Ruby is red transparent variety of corundum (gem.) It is found in the areas of Zanskar, Sumjam and Poddar in association with sapphire bearing rocks. Ruby occurs in metasediments and meta-basic intruded rocks such as quartz and pegmatites. A highly productive locality for aquamarines was discovered in the Kashmir Valley in Skardu, whence crystals of considerable size and purity were recovered. The gem occurs in coarse pegmatite veins traversing biotitegneiss.

Tourmalines

Pellucid and beautifully coloured varieties of tourmaline, red, green or blue, are worked as gems. The green variety known as indicolite occurs in the Poddar area of Kashmir where also some transparent crystals of rubellite are found. The latter tourmalines possess greater transparency, but are much fissured. The Geological Survey of India is exploring new sites in the Doda district of Jammu to discover tourmalines.

Besides the above named varieties, other crystallized minerals, when of fine colour and attractive appearance and possessing some of the other qualities of gams, e.g., hardness, transparency, etc., are cut for ornamental purposes in Kishtwar, Doda, Kashmir, Ladakh. Turquoise, opaque, of fine blue colour, usually uncut which is commonly sold in Srinagar is a product of Tibet, or Ladakh. Gem-cutting industry has great potential in Kishtwar, Srinagar, Jammu, Leh and Kargil.

There are numerous other non-metallic minerals found in the state, most of which are bedded deposits.

Bentonite

Bentonite is especial assemblage of clay mineral, in many instances formed by weathering of acid, lavas and pyroclastic rocks. The deposits of bentonite are spread between Bhimber and Kathua in the Jammu Division. These deposits lie, on the gentle dipping slopes of the outer plains of Jammu. Para, Chittopali, Modkali, Uttar Bhimber Bazar and Panipur are the areas of Jammu Division in which bentonite is found. From Bhimber to Kathua there is a continuous bed of bentonite deposit running for a length of about 10 km and about half a metre thick occupying a definite horizon in the Upper Siwaliks.[8]

Bentonite is used for cleaning grease spots, soap making, ceramics plaster, and filling paper, washing silk and cotton clothes. It is also known as '*Serati-Mitti*'.

Kaolinite (China Clay)

Kaolinite is a clay produced by the decomposition in-situ of the felspar. It is the natural (unwashed) china clay. Kaolinite is found in association with bauxite deposits of Jammu. It occurs between great limestone and breccia and nummulitics and coal measures of bauxite series of the Salal, Songarmar and Chakar areas. The local name of the mineral is *Makol*. It is used for white-washing purposes. Kaolinite deposits are available upto a thickness of one to four metres in the areas of its occurrence which can be used effectively in pottery making.

Borax

Borax is deposited as a result of evaporation of hot spring water. It is of use in the manufacture of superior grades of glass, artificial gems, soaps, varnishes and soldering and enameling. Ladakh is the only part of the country in which borax is found. In the form of surface deposits, borax is found in the Puga Valley and Chumathang of the Maheyang area of Ladakh. Annually about 15 tonnes of borax is obtained from these places. There is a project to refine the large reserves of crude borax in the Puga Valley lakes locally and to transport the product by aircraft to the industrial centers of India.

Gypsum

Gypsum forms large bedded masses or aggregates occurring in association with rocks of a number of different geological formations. Millions of tonnes of gypsum, the alteration product of pyritous limestone of Salkhala age, are laid bare in the mountains of Uri and Baramulla area of Kashmir in a stretch of about 40 km along the strike.[9]

Gypsum is an evaporite mineral found in clays and limestones, sometimes associated with sulphur. It is an important non-metallic mineral of great economic importance. It is a hydrous sulphate of calcium. Snow-white gypsum with intercalation of quartzite and schists is found in areas of Uri, Ramban, Assar, Batote. The industrial use of gypsum is in the manufacture of synthetic fertilizers, plasters, distempers and in the cement factory. It is also used as a surface-dressing for lands in agriculture, and as a fertilizer, with considerable benefit to certain crops. Gypsum has begun to be used as a source of sulphur in the manufacture of fertilizers. Gypsum mined near Uri is supplied to the Wuyan Cement Factory of Srinagar (Kashmir). Ramban and Assar fields have gypsum deposits of about 19 and 5 million tonnes respectively.

Magnesite

Magnesite is a mineral of the carbonate group. It is found in irregular veins in serpentine and formed by replacement of dolomite and limestone. In the state of Jammu and Kashmir, magnesite is found in Shergol (Leh-Kargil Road). The recent researches by Rana (1975), Chadha (1978) have discovered good deposits of magnesite in and around Katra (Udhampur) area of the Jammu Division.

Sulphur

Sulphur is a mineral obtained form iron pyrite or in the neighbourhood of volcanic regions. It is also associated with limestones, sandstones and saltdomes. In the state of Jammu and Kishmir, sulphur is found in areas of recent volcanic activity, around hot springs. Substantial deposits of sulphur have been located in the Puga Valley of Ladakh Division. The Geological Survey of India has estimated a reserve of about 6,000 tonnes of sulphur in the Puga Valley.

Limestone

Limestone is a carbonate of calcium. Besides, its uses as building stones, and as lime and cement raw material, limestones, if of the required purity, have important uses in the chemical, alkaline, sugar and metallurgical industries.

It is found in Manasbal, Beru, Achabal, Barwar, Verinag, Wuyan, Jangalgali, Salal, Riasi, Basohli, Kalakote and Khalsi (Ladakh).

Slates and Phyllitics

Slates and phyllites are the metamorphic rocks of great economic importance. Slate is obtained form the vicinity of Banihal, Ramban and Ramsu. It is being used successfully for roofing. In Marpathri near Gulmarg about ten million tonnes of good quality of roofing slate has been estimated by the Geological Survey of India.

Mineral Paints

A number of rock and mineral substances are employed in the manufacturing of paints and colouring materials. Substances which are suitable for this purpose include earthy forms of hematite and limonite (ochre, *geru*), refuse of slate and shale quarries, possessing the proper colour and degree of fineness; graphite; laterite, orpiment; barytes, asbestos, mica, steatite, etc. Large quantities of red and yellow ochre in association with graphite-bearing slate occur in the Salkhala System of deposits in the Uri Tehsil of the Baramulla District of Kashmir Division.[10]

Sand and *Bajri*

Enormous quantities of sand, pebbles and shingles are available in the Chenab and Tawi rivers and the numerous Nallas and hilly-torrents of the state. Sandstones are in abundance along the Jammu-Srinagar National Highway and the Jammu Punch Road in the Siwaliks and Murree tracts which are being used for building materials.

Mineral resources as discussed are exhaustible and therefore, should be used judiciously. The exploitation and conservation of mineral resources largely depends on the technological stage, the location of mineral ore with reference to transport and market. The exhaustible mineral resources deserve special attention. A scientific policy about their exploitation and utilization needs to be formulated.[11]

Minerals Conservation and Environmental Protection

Extraction of minerals is our economic necessity. Our economic structure which is quite complex hinges on industries. These industries draw their raw materials from minerals, forests and agriculture. The large-scale and heavy industries, economic and commercial activities are largely dependent on minerals and metals. The process of extraction of mineral resources and its use in various ways generate a wide range of environmental changes—sometimes having far reaching consequences. While these wasting assets have evoked formulation of national policies for their proper conservation, little has been done to ensure rehabilitation of land adversely affected by mining activities. Since we can not sustain our economy and society without industrial development, it is imperative to utilize the resources judiciously and to manage the mining industry scientifically. The damage caused to the environment may be reduced if the following steps are taken by the government, the contractors and the people who are directly or indirectly involved in the mining industry.

1. There should be a legislative provision for full responsibility of treatment and environmental rehabilitation on excavated areas squarely on the mining agencies. There should be legal binding on the mine owners to restore the land they have damaged.
2. There must be compulsion for stacking and preserving topsoil in order to cover back the excavated parts with debris produced and then put a cover of soil over it so that crops and grasses may be grown over it.
3. There should be incentives for successful programmes of land restoration, soil conservation and afforestation.
4. There should be serious attempts to stabilize the slopes, not only during the excavations but also afterwards to ensure that stability is maintained in future also. For this purpose, there should be construction of drainage network, and retaining walls.
5. There should be restriction on mining in the geo-dynamically sensitive and ecologically fragile parts of the mountains.
6. Deeper excavation on the surface or underground digging, causes the water table to sink locally, resulting in the wells and springs of the vicinity going dry. Mining of limestone, etc., in the neighbourhood of towns and villages therefore should be restricted.
7. The miners should be provided safe drinking water by the mine-owners.

8. The deep depressions formed by deep mining can be converted into lake to develop picnic spots or they may be used for dumping municipal and industrial wastes. These waste dumps may be subsequently covered with topsoil and afforested.
9. There should be more research to develop the new technology for the optimal extraction of minerals which may be eco-friendly.
10. There should be environmental awareness among the masses who can help in the restoration of the degraded mining tracts.

Notes

1. Whitten, D.G.A. and J.R.V. Brooks, 1977, *The Penguin Dictionary of Geology*, Penguin Books, New York.
2. Wadia, D.N., 1994, *Geology of India*, the 4th edn. Tata McGraw-Hill, New Delhi.
3. Srivastava, B.S., Nanda, H.M., 1976, Coal Resources of Jammu Area—An Appraisal, *Abstracts ASI*, p. 39.
4. Pareek, H.S., 1976, On the Study of Solid Fuels of Jammu and Kashmir State, *Abstracts Geological Survey of India*, pp. 37-38.
5. Middlemiss, *Mineral Survey Report*, Jammu and Kashmir State, Coalfields of Riasi, Jammu, 1930.
6. L.A.N. Iyer, *Indian Precious Stones*, Calcutta, 1946.
7. Wadia, D.N., 1994, op. cit. pp. 467-478
8. Ibid.
9. C.S. Middlemiss, *Mineral Survey Report Jammu and Kashmir State* (Graphite and Ochre), Jammu 1926.
10. *Digest of Statistics, Jammu and Kashmir State* 1994-95.
11. Mohammad A., 1986, Jammu Division, Regional Structure and Levels of Development, Unpublished Ph.D. Thesis, Kashmir University, Srinagar.

6

Ethnic Composition

An ethnic group is a distinct category of population in a larger society whose culture is usually different from that of the society. The members of such a group are, or feel themselves, or are thought to be, bound together by common ties of race or nationality or culture. In other words, a group of people having common racial, religious, linguistic or national characteristics is known as an ethnic patterns, business success and the political power base[1].

The history of peopling to Jammu and Kashmir state is a record of constant impulses of immigration form the north-west, west, south and south-east directions. The alien races, ethnic groups and various religions and cultures have influenced the cultural ethos and lifestyle of people of this region[2]. It is however, generally accepted that upto the beginning of the 14th century, the state of Jammu and Kashmir in general and the valley of Kashmir in particular was inhabited by Hindus, and that by the end of the 14th century the mass of people embraced Islam. In the Valley of Kashmir, it was because of the preaching of Shah-i-Hamadan that most of the people followed the teachings of Islam[3].

The ethnic mosaic of the Jammu and Kashmir state is quite complex. The racial and ethnic structure can not be explained without understanding the pre-historic movements of the people. In the process of peopling of the region, the Dards in the north-west, the Ladakh is in the north-east, the Gujjars and Rajputs in the south and the *Paharis* in the south-east have closely influenced the existing ethnicity of the people of the region. The racial composition of the state has also been influenced by the immigrants from the territories of Afghanistan. Azerbaijan, Georgia (Gurgia), Iran, Iraq, Kazakistan, Syria, Sinkiang (China), Tadzkistan, Tibet, Turkmenistan, and Uzbaikistan[4].

The different races, ethnic groups and castes of the Jammu and Kashmir state, though intermingled, have their areas of high concentration. For example, the Kashmiris—an amalgamation of numerous cultural groups—are mainly concentrated in the Kashmir Valley-floor (Jhelum Plain); the Dards occupy the Valley of Gurez; the Hanjis are confined to the lakes, rivers and water bodies of Kashmir; the Gujjars and Bakarwals (goat-keepers) are occupying the kandi areas and undulating slopes; the Dogras have their dominance at the outskirts of Punjab plain; while the Chibalis (Chibs) and the Paharis live mainly between the Chenab and the Jhelum rivers. Moreover, there are numerous small ethnic groups like Bhotias, Gaddis, Sikhs, Buddhists and Jains who, in isolated pockets, have significant concentration.

Kashmiris

'Kashmiri' is a wider term. It consists of numerous ethnic and cultural groups. Though living and scattered in different parts of the state, their major concentration lies in the Valley of Kashmir, Bhadarwah, Doda, Kishtwar, Banihal, and Ramban Tehsils of Jammu Division. They are also found in Kargil, Leh (Ladakh), Rajauri, Poonch, and Udhampur districts of the Jammu Division.

The word 'Kashmiri' has been loosely applied for several streams of immigrants, mainly from Afghanistan, Iran, Turkey and Central Asia, There is a close bearing of the Indo-Aryan culture on the ethnic composition of Kashmiris. In fact, the Kashmir language has its origin from the Proto-Indo-European family, also known as the Kurgan language. Moreover, Kashmir has also received racial impulses from the Indo-Greeks which have influenced the ethnic structure of Kashmiris significantly. The influence of Dards, Ladakhis, Punjabis, Sikhs, Rajputs, Gujjars has also moulded the cultural ethos of the Kashmiris[5].

Kashmiris are generally broad-shouldered and usually medium to tall stature. They are much dolichocephalic, having a well-developed forehead, a long narrow face, regular features and prominent straight, finely cut nose, attractive personality and graceful figures. They have keen trading instinct, and are deeply attached to their country. A large number of unemployed Kashmiris will not seek jobs outside of Kashmir. They wear short *pyjamas*, a long loose large-sleeved gown, locally known as *pheran* and a skull cap. In intellect, they are considered to be superior to their neighbours. In disposition they are talkative, jolly and humorous[6].

About 95 per cent of the total population of Kashmir consists of Muslims. Every Kashmiri Hindu, with the exception of Khatri shopkeepers of Srinagar, who are known as *Bohras*, is called a Pandit or learned Brahman. Though not all learned, they are all Brahmins, and chiefly the followers of Shiva, the Lord of the Mountain and the God of the hill people. Khir-Bhawani at the mouth of the Sindh Valley is their favourite goddess.

The Pandits divide themselves into three classes: (i) the astrologer class (*Jotish*), the priest class (*Guru* or *Bachabat*), and the working class (*Karkun*).

The priest class generally do not marry with any of the other classes, partly because they are regarded as divine and cut off from mankind, and partly because the laity abhor their practice of accepting the apparel of deceased Hindus. But the *Jotish* and *Karkun* pandits intermarry[7].

The Pandits are broken up into numerous *Gotras*. There are 18 *Gotras* among the Shivites Brahmans and 103 among the other Brahmans in Kashmir. Marriage is forbidden within the *Gotra*. The Leading *Gotras* are: Tikku, Razdan, Kak, Munshi, Mathu, Kachru, Pandit, Sapru, Bhan, Zutshi, Raina, Dhar, Fotedar, Madan, Thusu, Wangnu, Muju, Hokku, and Dulu[8].

The Kashmiri Muslims are predominantly cultivators, living in rural areas except those who are in urban places like Srinagar, Baramulla, Anantnag, Sopore, Pulwama, etc. Among the Muslims, the largest number consists of Sheikhs. The Kashmir Muslims, by and large, are the descendants of original Hindus. The Saiyes, Mughals, and Pathans are the other major Muslim ethnic groups.

The Muslims of the valley may have retained, some of the customs of endogamy within the caste and exogamy outside the *Gotra*. But there is no trace now of these customs, and only the different *Gotra* names are names and nothing more.[9] It would be interesting to trace the origin of the *Gotra* (*Kram*) names, although by intermarriage the *Krams* have ceased to have any individuality or distinction. It is supposed by many that Muslims of the Pandit, Kol, Butt, Aitu, Rishi, Mantu, and Ghanai *Krams* are the descendants of Brahmans. Other *Krams* are believed to have sprung from *Khatri* origin. The Magres, Tantres, Dars, Dangers, Raina, Rathors, Thakurs and Naiks are from the Khatri *Krams*. Only one *Kram*, the *Lone*, is generally assigned a Vaisya origin, and the *Damras* are said to be the descendants of Sudras, the lowest of the Hindu castes[10].

In recent years, especially after 1947, it has become difficult to distinguish between the high and the low *Krams* as the men of low occupations are arrogating high sounding names. Thus of late years the *Dums* of Kashmir have steadily assumed the *Kram* of *Ghanai*, much to the annoyance of the original *Ganar*. The Hanjis (boatmen) of Kashmir have assumed the name *Dar* as a patent of respectability. Some *Krams* are, however, restricted to men of-menial pursuits and the *Kram* name *Sufi*, which is said by some to be the Brahman origin is mainly found among market-gardeners, bakers and servants[11].

Among the Shaikhs, the Pirzadas, who are the descendants of zealous converts of Islam, consider themselves to be the Saiyeds and intermarry with them. The Babas, also descended from zealous converts, are now chiefly religious mendicants. The Rishis are the attendants at shrines. The *Mullas* or priesthood of Kashmir are Sheikhs, and may be divided into two classes. The first includes *Mullas* learned in the law, and variously designated as *Maulvi*, *Qazi*, *Akhund* or *Mufti*. The second class consists of *Mullas* who have fallen in social position and are known as *Mals*. These wash and prepare the bodies of the dead for burial and dig grave. Many Dums and Hanjis have adopted *Kram 'Mal'* but the Hanjis is regard the name as a corruption of the Hindi word for boatmen (Mallah).

The Saiyeds constitute the other important ethnic group in Kashmir. They may be divided into those who follow the profession of *Piri-Muridi* and those who have taken to agriculture and other pursuits. Some Saiyed families are highly respected in the villages, but those who have taken to agriculture are practically on a level with other villagers and intermarry with them[12].

Mir is the *Kram* name of the Saiyeds, while he retains his saintly profession, the Mir is prefixed to his name, when he takes to mundane pursuits the Mir is suffixed to his name.

There are not many Mughals in Kashmir, and have so intermarried with the ordinary Kashmiri Muslims that all trace of descent is lost. They came to Kashmir during the period of Mughals. Their *Krams* are Mir (a corruption of Mirza), Beg, Bande, Bach, and Ashaye.

In comparison to Mughals, the number of Pathans is large. Many of the Pathans retain all the old Pathan customs and still speak Pashtu language. They pride themselves on their bravery. They have large concentration in Uttar Machipura, Haval (Hamal). Most of the Pathans came to Kashmir in the time of Durrani. The Pathans are always given the title of Khan.

The Bombas are believed to have immigrated in Kashmir from Turkey. They are chiefly confined to Machipura. Originally a warlike tribe, they have degenerated into a feeble and most pitiable condition. They intermarry among themselves. Batkot is the old home of the Bombas in Kashmir, and when they die they are brought back to the beautiful burial ground, still kept up in Batkot. The heads of Bomba families are addressed as Raja, and the tract in which they live is known as *Rajwara*[13].

There are several villages of Faqirs or professional beggars. They work as agriculturist during the summer season and beg during the winter season. They regard the profession of begging as most honourable. They are not aggressive in their calling and are liked by the people.

In addition to these, there are numerous inferior castes who assumed their *krams* from the occupations they are engaged in. The market gardeners, herdsmen, shepherds, boatmen, minstrels, cobblers, leather workers, and scavengers. These small castes are scattered all over the state.

The Dums of Kashmir is an important caste, and have upto quite recent times had great power in the villages. The village watchman was always a

Dum, and in addition to his police functions he was entrusted by the state with the duty of looking after the crops. The Dums, like Galawan and Batal generally, have darker complexion[14]. The Dums are much dreaded and disliked. Many stories are told by the Dums about their origin, and they always claim descent from a Hindu king, who afraid of his numerous sons, scattered them all over the valley, but it is believed that they are of *Sudra* origin.

The Galawans, or horse-keepers of Kashmir are by some considered to be the descendants of the Dums, and their dark complexion suggests that they are not of the same race as the Kashmiri peasants. Others think that the Galawans are the descendants of the *Tsak* Tribe. Originally they used to earn their livelihood by grazing ponies but with the advent of time they found it more remunerative to steal them and they eventually developed into an established criminal tribe. They achieved notoriety in the days of Pathan rule, and when the Sikhs took over Kashmir, the Galawans were a terror to the country. They used to move about in large parties, all mounted and armed with long, heavy clubs. They used to raid threshing-floors and frequently attacked a wedding party, and carried off the bride. Khaira Galawan is the hero of many a legends. The term 'Galawan' is now used to describe a man of violent and predatory habits.

The shepherds of Kashmir are known as Chaupans or Pohls. They are considered to be a separate ethnic group. A Chaupan is generally a cheery, active man with a most characteristic whistle, and his healthy life in the high mountains makes him strong and robust. They occupy different *Margs* (alpine pastures) in which they graze their goats and horses during the summer season. In the autumn season they descend in the lower altitudes and live in villages.

Band or Bhaggat combine the profession of singing and acting with that of begging and are great wanderers. They are much in request at marriage feasts and at harvest time they move about the villages. They sing songs in Kashmiri, Persian, Panjabi, Urdu and Hindi. They are very pleasant people[15].

Hanjis

Hanjis are essentially the dewellers of water and are mainly confined to the Dal, Wular, Anchar and Manasbal lakes and the Jhelum river especially between the Khanabal District Anantnag and Chattabal (Dist. Srinagar). They are an important ethnic group in the Valley of Kashmir. It is believed that Raja Parbat Sen introduced boatmen in Kashmir from Sangaldip. They were of the *Vaisya* caste, and even now some of them are termed as *Sudra*. Some of them believe as they are the descendants of Hazart Noah[16]. Among Hanjis, the father of the family is an autocrat. Hanjis are active hardworking people and small children start the work of towing or paddling at a very early age. On the basis of economic activity and occupations, the Hanjis are divisible into several categories. The major sub-groups of Hanjis are: (i) Demb-Hanz (vegetable growers), (ii) Gari-Hanz (waternut gatherer/s), (iii) Gad-Hanz (fisherman), (iv) Mata-Hanz (who deals in wood), (v) Dunga-Hanz (owner of passenger boats), (vi) Haka-Hanz (gatherers of wood from water bodies), (vii) Bahatchi-Hanz (who live in Bahatch boats), Shikare Hanj (who ply shikara boats), and (ix) Houseboat-Hanz. The various type of Hanjis use boats of different shapes and sizes. The type of boat which a Hanji used denotes his caste, social status and occupation. Their favourite *Krams* are Danger, Dar, and Mab.

A field survey conducted in the Dal Lake in 1999 shows that the sex ratio of Hanjis is about 892 against 927 (1991) for the country as a whole. The low sex ratio shows that the females are not properly nourished and provided medical and educational facilities. In fact, in the male dominated society of Hanjis the females are generally the ignored lot.

Among the Hanjis the male and female literacy rate is 21 and 5 per cent respectively. Nearly 35 per cent of the Hanjis workforce is engaged in primary and tertiary activities, while the rest 65 per cent is the dependent population consisting of the juvenile and senile population. They do not have fixed income which varies from month to month and from season to season. Their income has declined substantially since 1989 after the militancy and political instability. They are largely dependent on tourism, vegetable cultivation, fisheries and water transportation.

Economically, the Hanjis are backward. The Gad-Hanj and the Demb-Hanj are maintaining a poor standard of living. The houseboat owners are, however, rich who are maintaining a good standard of living. Some special strategy needs to be designed to raise their income base which shall ultimately enhance their educational standard and standard of living.

Watals

The Watals have been called the gypsies (wanderers) of Kashmir. Socially, the watals have been divided into two groups: (i) those who abstain from eating carrion, and admitted to the mosque, (ii) those who eat the flesh of dead animals, and are excluded from the mosques. The principal occupation is the manufacture of leather. The watals of the first category make boots, and sandals, while the watals of the second class manufacture winnowing baskets/trays of leather and straw and perform the duties of scavengers. Watals settlements are always at some distance from the cottages of peasants in the villages. They prepare the hides of dead cattle and buffaloes and the skin of sheep and goats and keep poultry for sale. The outcast watals have no religion, but they respect the shrines and often visit them, though they are not allowed to enter the holy precincts.

Dards

Dards constitute a peculiar ethnic group. Their major concentration is in Dardistan (Dardesa), the land lying to the north of Kashmir Valley especially in the catchment of Kishanganga north of Sardi, Gurez and Tilel.

Dards have a long history, though the period of their arrival in this area is not known. Ptolemy in his book *Almagast* has used the word Daradari for Dards of the Western Himalayas. In the opinion of Leitner, Dards belong to the Aryan stock. This opinion is also endorsed by Ray who states that the Dardic Aryans Parted from the main Aryan mass just after their entrance into India. Dardic Aryans then colonized the Pamir region from where they spread to Chitral and Gilgit. Before embracing Islam, they were the followers of Buddhism and Hinduism.

In physical appearance, the Dards are broad-shouldered, moderately stout-built and have well-proportionate bodies. In face, they are not handsome, their heir is usually black but sometimes brown. In complexion, they are moderately fair. Their eyes are either black or hazel. They are well-known for their ferocity. In the social hierarchy, they are divisible into: (i) Renue (ruling class), (ii) Shins (religious sect), (iii) Yashkun (cultivators and (iv) Dum (menial class). Most of the Dards are dependent on agriculture, forestry and cattle keeping. Their literacy rate is less than 20 per cent and the female literacy is only about 8 per cent. Being male dominated society, the status of females is low and they hardly have any idea about their fundamental rights and liberties.

Gujjar-Bakarwals

The Gujjars is a unique and significant ethnic group of Jammu and Kashmir. They are dependent largely on cattle, goat, sheep and horse keeping. The Gujjars are divisible into two groups: (i) those who practice transhumance and oscillate with their *Rewads* (herds) between the alpine pastures (*margs*) and the low altitudes of the Siwaliks and the Jammu plains, and (ii) the sedentary Gujjars who settled in most of the villages situated along he annual routes of migration[17]. These sedentary Gujjars claim a common and collateral ancestry.

There is not an unanimity among the social scientists about their origin and arrival in India. Some scholars are of the opinion that they are the foreign stock representing those pastoral nomads from the steppe grasslands of Central Asia, who entered India either with the Huns or a little later. Some of them are of the opinion that the Gujjars are the descendants of Kushan and the Yuchi tribes which are considered to be the tribes of Eastern Tatars (Russia). In the opinion of a sizable number of social scientists, archaeologists and linguistics, they are the descendants of Gurjis (Georgians) who inhabit a territory between the Black Sea and the Caspian Sea, south of the Caucasus Mountains. This opinion is based on the records and names of different villages, areas, settlements and towns. One of the major traditions of Gurji (Georgians) people right form the earliest times had been that they used to give their tribal names to the places and localities they inhabited. It is said that before their march to the Indian subcontinent, they occupied some places and localities in Iran, Iraq, Central Asia, and Afghanistan. These are known as Jurz, Gujur, Guzru, Gurjistan, Gujar-khas and Chusak-Gujar. When they came to India, they named certain areas as Gujarat, Gujranwala (a district of Punjab in Pakistan), Gujarat (India), Gujargarh (Gwalior), Gujarwada and Gujarat (Saharanpur and Hardwar districts of U.P.). Many other smaller places also have their names after the Gujjars[18].

If the above opinion is taken correct, then it can be said that the Gurjis (Georgians) under certain pull and push factors migrated from their original habitat to India through Iran, Central Asia and Afghanistan. They entered the Indian subcontinent through the passes of Hindukush, Sulaiman and Kirthar Ranges. Around the 4th and the 5th centuries A.D., they occupied the territory around Mount Abu, the Peninsula of Kathiawad (Gujarat) and the adjacent hilly tracts of the Aravallis. Later on facing serious drought conditions they migrated from the south-western parts of Rajasthan and occupied the plains of Punjab. Subsequently, they occupied the green pastures of the Siwaliks and the Himalayas in the state of Jammu and Kashmir. The Gujri language is now recognized to be a form of the Rajasthani language. Their language is also known as Parimu or Hindki is wholly different from the Kashmir and the Dogri languages. They rarely intermix with the Kashmiris, though like them, they are Musalmans by religion.

In appearance, the Gujjars are sharp-featured tall race. They are simple, less educated and inoffensive people. Their good faith is proverbial, and they are generous people.

On the basis of their occupation, the Gujjars may be divided into two groups: (i) cultivators who have sedentrized on the slopes and the side-valleys lie Naranag, Wangat and Daksum, and (ii) the Gujjars who practice transhumance. The second category is further sub-divided into *Baniharas* or Dudhi Gujjars (milkmen), and the *Bakarwals* (who rear sheep and goats). The *Banisars* or the Dudhi-Gujjars tend buffaloes, sell *Dudh* (milk) and milk products and for this reason they are known as *Dudhi* (milkmen). They live in *Bans* (forest). The Gujjars and Bakarwals keep herds of goats owing to which they are called as Bakarwals. The Bakarwals seasonally migrate from lower altitudes to the alpine pastures of the Himalayas in summer and descend to the lower altitudes in the winter season. The Banihari Gujjars who keep buffaloes move within a short distance of about 50 km from their base in a year. Some of the Baniharis and Bakarwals supplement their income from agriculture and forestry.

The pastoral economy of Gujjar-Bakarwals depends mainly on the availability and utilization of extensive seasonal pastures. While snow covers the alpine pastures (*Margs*) in the north, relatively poor pastures are available throughout the winter in the south. In spring season, the pastures are plentiful and good in the areas of low and middle altitudes, while the scorching heat of summers parchases the pastures in the south, good pastures are found in the north above 2700 m where melting of snow gives way to green pastures. Thus, both the winter and the summer zones are characterized by the availability of pasturage in a definite part of the year. This leads to the seasonal migration of the Bakarwals with their herds from the Siwaliks to the Middle and Greater Himalayas in summers and downward movement in winter season.

Transhumance (Seasonal Migration) of Gujjars and Bakarwals

The practice among pastoral communities of moving their herds or flock seasonally or periodically between two regions of differing climatic regions is known as transhumance. The practice is particularly characteristic of mountainous regions where animals are usually transferred from mountain to valley pastures in the winter and back again to the mountains for the summer,[19]. Such movements are always accompanied by herders.

Transhumance is a response of pastoralists to topographic and climatic constraints. Protection of their herds according to their climatic tolerance and optimization of the use of pastures form part of this response. This results in oscillation through time over space or in cycle mobility in a space-time continuum. The temporal scale of transhumance may vary from long-term cycle spread out over a number of years or an annual cycle or diurnal cycle. The term 'space' refers to different spatial zones of regions covered by a transhumance in a transhumance ecosystem. Although it is an economic necessity for the Gujjar-Bakarwals to move with their flocks in each season to the areas where pastures are available, yet the migration has a greater value than this. They have two ways of expressing the same experience. Time and space for them have the same connotation, as every destination of theirs is intimately linked up with a season (time) or every part of the year is linked up to a space which is their destination. This explains that they interpret time and space in the same terms with reference to migration.

FIGURE 6.1

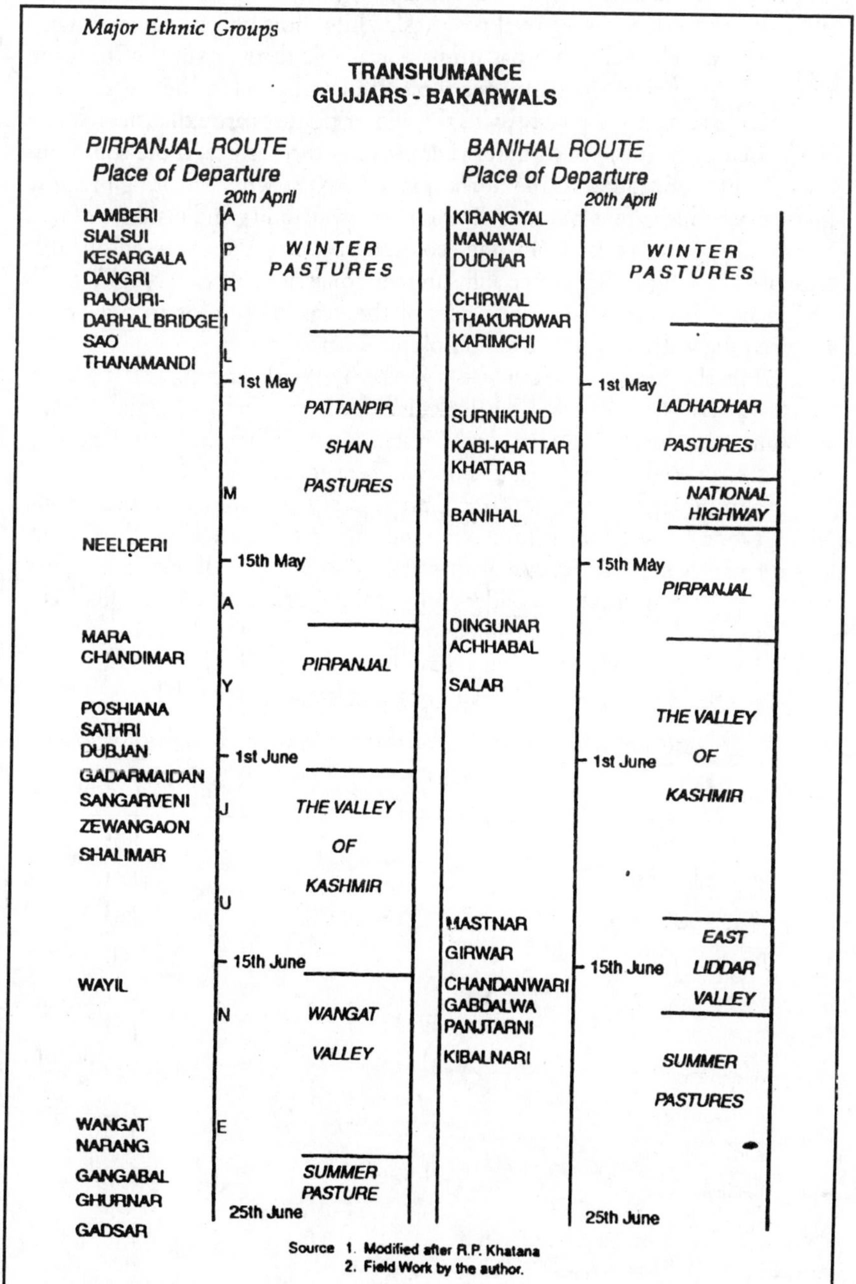

Source 1. Modified after R.P. Khatana
2. Field Work by the author.

The pastoral economy of Gujjar-Bakarwals depends on the validity and utilization of extensive seasonal pastures. While snow covers the mountains in the north, relatively poor pastures are available throughout the winter in the south. In spring, the pastures are plentiful and good in the areas of low and middle altitudes. By late April, the winter pastures are exhausted. While the scorching heat of the summer is desiccating the pasture in the south, usable pastures are found in the north above 2,800 m where melting of snow gives way to lush green pastures. The relative availability and lushness of pasturage, therefore, varies from one ecological zone to the other within this habitat. Thus, both the winter and summer zones are characterized by availability of pasturage in a definite part of the year. This leads to oscillation between the summer and the winter pasture zones.

While the advent of summer is signaled by the drying up of the pasturage in the south, they take the flock during this period to the pastures in high altitudes, locally snow as *dhoks*, in the north. When the high land pastures get covered under ice and snow and severe cold affects the zones of lower altitudes in the month of September. Every year they repeat this cycle of migration between the summer and the winter pastures. Thus, the Gujjar-Bakarwals organize their annual migration from winter abode to summer abode and back to meet the fodder and water requirements of their flocks.

The major oscillation channels of the Gujjar-Bakarwals and their altitudes through the Pir Panjal Range have been given in Table 6.1.

Table 6.1

Major Oscillation Channels of Gujjar-Bakarwals through Pir Panjal Range

	Osciflation Channel	*Altitude (in metres)*
1.	Jamiangali Pass	4088
2.	Toshmaidan Pass	4091
3.	Nurpurgali Pass	4093
4.	Pir Panjal Pass	3491
5.	Daṛhal Pass	3946
6.	Rupari Pass	4099
7.	Budhil Pass	4261
8.	Kausarmarg Pass	3123
9.	Didamgali Pass	3809
10.	Banihal Pass	3224
11.	Sarthal Pass	—

Source: After R.P. Khatana, p. 57.

All the tracks of the Gujjar-Bakarwals in the state of Jammu and Kashmir start from their winter bases situated between 600 and 1,200 m above the sea level, south of the towns of Rajauri, Riasi, Udhampur and Sambha. Although there are numerous passes in the Pir Panjal Range, they follow either the Pir Panjal route or the Banihal Pass route. A brief description of these routes and the intervening pastures and the problems they confront have been given in the following paragraphs.

The Pir Panjal Route

The *qafilas* of the Gujjar-Bakarwals who follow the Pir Panjal route start from their winter pastures in the third week of April and take about sixty days to reach the *dhoks* (summer pastures). Out of the 60 days, 22 are journey days and 38 are halt days. The *qafilas* start from Lamberi and making halts at suitable intervening pastures, they cross Sialsui, Kesargala, Dangri, Rajauri, Saj, Thana Mandi and Neelideri (Fig. 6.2). At Neelideri, they stay for about two weeks and this period of stay is utilized for grazing in the surrounding intervening pastures. At this halt they, purchase the required goods, give salt to their flock, and repair their tools and equipments.

From Neelideri, they move to Mara, Poshina, Chandimar and cross the stream at Chitta-Pani in which ice pieces flow in water. In this track, they face a lot of difficulties in fording the Chitta-Pani stream. At the occurrence of hails and rains, the flock suffer great hardship.

From Poshina they move to Gadarmaidan, crossing the Mara Chandimar, Posluana and Sathri Dubjan. It is in this tract where they cross the Pir Panjal Pass. Subsequently, from the Gadsar Maidan, they move to Sangerveni and then to Shalimar, situated to the north-east of Srinagar City on the bank of Dal Lake. After staying for about two or three days at Shalimar, they commence their onward journey towards Wayil, Kangan, Wangat and Naranag. From Naranag there is a steep ascend towards Gadsar. In fact, Gadsar pasture is the summer base for this *qafila*. During the summer season, they graze their sheep and goats in the *dhoks* (summer pastures) situated around Gadsar.

The Banihal Route

The second important channel through which the Gujjar-Bakarwals migrate from winter to summer pastures and back is the Banihal Pass route. Around 20 April, they commence their journey form Kirangyal and its environs and up to the *dhoks* (summer pastures) they break journeys at twelve places. Friday or Monday are considered as the auspicious days for the upward or downward migration. Starting from Kirangyal they break their journeys at Manawal, Katta (Ladhadhar intervening pasture), Surinkund, Kabikhattar

FIGURE 6.2

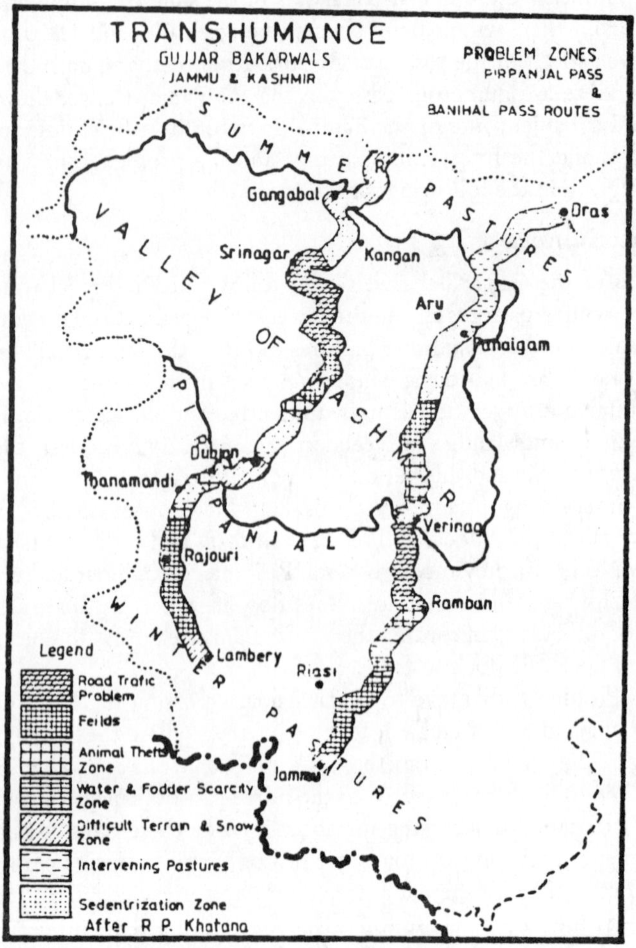

Banihal, Dingunar, Achabal, Salar, Masitnar, Girwar, Chandanwari, Gabdalwa, Panjtarni and Sukhnala.

The journey between Khattar and Banihal is full of anxiety and tension. Between these two places, the flock is to travel on the National Highway 1. This is a continuous 23-hours journey. Being a very busy route, there is great fear of accidents, lack of fodder, and animal lifters. This journey is broken at the Shaitani-Nala (near Banihal) for one day rest after a long and hazardous journey. After crossing the Banihal tunnel, the *qafilas* with their flock enter the Valley of Kashmir. In their ascend they pass through Verinag, Aishmuqam, and finally reach the pastures of Daksum, Chandanwari and Sukhnala.

The return journey for the winter pastures (Siwaliks of Jammu Division) commences in the month of September. It is the period when night temperature in the summer pastures reaches close to the freezing point and the frequency of fog increases. The green pastures die out under the impact of frost and freezing night temperatures. In such a situation the herds need protection from severe cold of the high altitudinal pastures. Consequently, the herdsmen start descending from the summer pastures. The return journey also consumes about sixty days which is broken at traditional *paraos* (halts).

Despite hazardous journeys, transhumance is a necessity for the Gujjar-Bakarwals. If these transhumans were to spent summers down on he outer hills of the Jammu region, the sheep and goats would die, if not of heat at temperatures of 40°C, then of starvation. And if they were to attempt to spend winters in Gurez, Tilel, Matyana, Rangdoom, and Waidwan on greater Himalayas—4,500 m above the sea level—they would simply be buried under several feet of ice and snow.

Economy and Society

The Gujar-Bakarwals have divided themselves into three principal kinship groups: (i) the *dera* (household), (ii) *dada-Porte* (lineage), and the *gotra* (clan). The *dera* is the basic unit of social structure among the Gujjar-Bakarwals. They count their numbers and describe their grazing and *qafila* groups in terms of the number of *deras*[20].

A *dera* usually comes into existence when a person establishes an independent household, which happens normally after his marriage. Each son, thus, establishes his own *dera* as he gets married. A *dera* usually consists of five to six members. There is a division of labour among the members of the household on the basis of sex and age. Females perform the domestic tasks of cooking, washing, fetching of water, upbringing of children, collection of wood, and spinning and making of woolen garments. On the other hand, males perform more arduous tasks like herding of flock and cattle, repairing of tools and equipment, collection of grass, herbs, deer-musk, hunting of wild animals, ploughing and harvesting of crops. The household is, thus, a primary economic unit. A nuclear family is the production and consumption unit. A joint family which is generally large cannot survive on the meagre pasture resource as the transhumance are on the move for about 110 to 130 days in a year. The elders want the eligible married youngsters to shoulder the responsibility of looking after the flock of sheep and goats independently. This would ensure greater security to the herds and sharing all responsible jobs by the adult members.

Several *deras* (households) constitute a lineage (*dada-potra*). The pastures are allotted to the lineage and not to the individuals. In a lineage, there may be about two hundred persons. Usually, a Gujjar-Bakarwal father divides his property (animal wealth) among his male children as and when they get married. The lineage, thus, consists of several generations and includes cousins and distant relatives. They share the common pastures. The lineage unit is quite powerful administrative unit. Each lineage has a head who is responsible for the socio-economic and political activities of his group. The headship of the unit is based on the principle of primogeniture and passes on to the eldest son either at the death of the father or when the father opts to retire. Many political and administrative decisions are taken by the head and the members of the group follow these decisions unquestionably. The unit head also gives punishment to the wrong doers[21].

The entire Gujjar-Bakarwal community is divided into a number of *gotras* (clans). The members of a clan are the descendants of a common ancestor. The gotra system has been borrowed from their Hindu Gujjars. The gotra name is often suffixed to their names.

The transhumant mode of life makes it necessary to form a number of functional groups for the efficient pursuit of their pastoral activities. For most part of their lives the transhumants live as members of these functional groups.

The primary functional group among the Gujjar-Bakarwals is the herding unit. A herding unit is based on several *deras* (households) which is helpful in the efficient grazing and care of flocks. The size of the flock determines the size of the herding unit. If the size of flock becomes larger it affects the efficiency of grazing and the health of the flock. In such cases, some of the *deras* prefer to break away from the herding unit, and when the size of flock is reduced considerably, it joins the original herding unit.

The *qafila* is the second important functional group among the Gujjar-Bakarwals. A *qafila* is a group of families which move together during the seasonal migration and submit to the authority of *qafila* leader at least for the period of annual migratory cycle. It is the transhumant counterpart of the compact village settlement of sedentary population group. They form the *qafilas* to assist each other during the arduous journeys. The *qafila* among the Gujjar-Bakarwals is, however, not a permanent grouping nor does it enjoy equal importance and relevance during the entire period of the annual migratory cycle.

The Gujjar-Bakarwals have an established institution of *Zirga* (panchayat). The *Zirga* decides the disputes among its members.

By faith the Gujjar-Bakarwals are the followers of Islam. They accept and observe the basic principles of Islam. Within the limits defined by the general tenets of Islam, they are free to develop and elaborate their ceremonies and customs as an antonymous folk system.

Since the Gujjar-Bakarwals are on the move for most of the year, their social and cultural traditions are strongly influenced by the migratory pattern. The basic activities involve camping, herding, and travelling. The beliefs, rituals, and cerements of these people need to be considered in the context of these meanings in the cycle of migration which dominates the life.

The main religious activities and festivals observed by the Gujjar-Bakarwals are five times prayers, fasting in the month of Ramzan (Ramadan). Idul-Fitr, Idul-Azha, Naoroz, and Baisakhi. They start their upward journey after the celebration of Baisakhi festival in April. During the course of annual migration, they pass in succession many shrines and the graves of *Pirs* (holymen).

The sex ratio of the Gujjar-Bakarwals is highly eschewed. There are approximately 856 women per 1,000 men among the Gujjar-Bakarwals. This demographic imbalance has a direct bearing upon their marriage pattern. The ceremonies connected with marriage have been directly influenced by their mode of life. Among the well-to-do Gujjar-Bakarwals, the *Mangni* (engagement ceremony) is generally held at an early age of about eight years. The *Nikah* (marriage) usually takes place after five years from the date of engagement. Marriages generally take place during the summer season when everyone has more time to spare. Marriage among these people is a simple ceremony. The bridegroom's father is required to fix the *Mehr* (dower) at the time of marriage in the terms of animals (sheep and goats), jewellery, money and other property.

The marriage takes place according to Islamic procedure. The consummation of marriage, however, takes place after a ceremony called *Rukhsati* which is usually held three or four years after the *Nikah*. The cases of divorce are rare. The widow can remarry.

The Gujjar-Bakarwals bury their dead according to the Islamic rites. If a death occurs during migration, the dead is buried somewhere along the route. They heap up stones on the grave and every year as they pass through the route they pay respect to the departed soul and light a lamp on the grave.

In brief, the Gujjar-Bakarwals have an organized social life. The existence of social and economic institutions, functional groups, and social stratification, the customs, traditions and are the outcome of the transhumance. These innocent people are of the superstitious nature. Their innocent nature is exploited by the *Pirs*. The geo-ecological environment of their habitat has exerted a determining influence in shaping their mode of life, their cultural ethos, values, customs and traditions.

The arduous journeys performed by the Gujjar-Bakarwals, though help in the utilization of winter and summers pastures are, however, not conducive for the socio-economic development of these people. Sedentarization in the side valleys and at other suitable places, therefore, need to be done. For the sedentarization, it must be borne in mind by the policy-makers to identify the places and sites for which they are more attached and where they can sustain themselves by cultivation of crops and by herding[22]. These places where permanent settlement should be such where the quality of life is fully satisfying for the nomads. If sedentarization of the Gujjar-Bakarwals could be made at the places where some land for cultivation is available and pastures for flock are also in the vicinity of settlement, their quality of life may improve and the ecosystems may be more sustainable.

Dogras

On the outskirts of the Siwaliks facing the plain of Punjab is the habitat of Dogras a distinctive 'ethnic group of Jammu Division. There is controversy among the social anthropologists about their origin. The major concentration of Dogras however, occurs between the two holy lakes, i.e., Saroinsar and Mansar. Lake Saroinsar is at a distance of 38 km to the east and Mansar 64 km to the west of Jammu city.

In the opinion of some social anthropologists Dogra is a corruption of the Rajasthani word 'Dungra' meaning hill and when the people of Rajasthan migrated in the hilly tracts under drought conditions, the Rajputs gave this name to the people of hilly country. Stein opines that the name 'Durgara' is probably a tribal designation like 'Gurjara', original of the modern Gujjar, and similarly the word 'Durgara' has been derived form 'Durgar' through Prakrit Dogar[23].

Whatsoever the controversy about the origin of the word Dogra be, they belong to the Aryan race and speak the Dogri language. Most of them have Brahmanical path and have the sects of *Varnashram*. A substantial section of the Dogras embraced Islam during the 16th and 17th centuries. At the time of partition of the subcontinent, most of the Muslim Dogras migrated to Pakistan.

In appearance, Dogras are short statured, slim and have high shoulders. Their complexion is wheatish, slightly hooked nose, brown eyes and jet-black hair. The lower castes of Dogras in general have blackish complexion.

Notes

1. R.J. Johnston, 1986, *Dictionary of Human Geography*, Blackwell, Oxford.
2. H.N. Koul, 1958, A Survey of the Origin of People of Kashmir, *Kashmir Today*, vol. II, 9.p. 18.
3. P.N.K. Bamzai, 1962, *History of Kashmir*, Delhi.
4. M. Husain, 1986, *Geography of Jammu and Kashmir*, Rajesh Pub. New Delhi.
5. V. Kumar, 1968, *The Nilmat Purana*, vol. I, Jammu.
6. C.E. Bates, *A Gazetteer of Kashmir.*
7. M.A. Stein, 1979, *Kalhan's Rajtarangani.*
8. Ibid.
9. G.M.D. Sufi, 1979, *Islamic Culture in Kashmir*, New delhi.
10. S.C. Ray, 1970, *Early History and Culture of Kashmir*, Delhi.
11. F. Drew, 1971, *The Territories of Jammu and Kashmir*, Delhi.
12. L. Rookh Rub, *Race and Tribes----Culture of Kashmir*, New Delhi.
13. Ibid.
14. W.R. Lawrance, 1967, *The Valley of Kashmir*, Srinagar.
15. *Imperial Gazetteer of India*, vol. Xv, 1908, London.
16. A. Mann, 1984, The Hanjis of Dal Lake, unpublished M.Phil Dissertation, University of Kashmir.
17. R.P. Khatana, 1992, *Tribal Migration in Himalayan Frontiers,* Vintage Books, Gurgaon (Haryana).
18. Ibid. pp. 158-175.
19. Ibid. pp. 159-160.
20. Ibid. pp. 160-165.
21. M.L. Lidhoo, 1985, Gujjars—Their Attitude Towards Social Changes, *Education and Modernisation*, pp. 2-21.
22. M. Husain, 1986, op. cit.
23. J. Pathik, 1980, *Cultural Heritage of Dogras*, New Delhi.

7

Population and Demographic Profile

The spatial distribution of population in Jammu and Kashmir is highly uneven. The distribution and density of population differ from district to district and from micro to meso-regions. The physical factors like terrain, topography, slope, climate (temperature and precipitation), drainage soil, natural vegetation, mineral resources and accessibility have largely controlled the distributional pattern and density of population in the state. In general, about 85 per cent of the total population of the state is occupying the Jammu Plain and the Kashmir Valley (Jhelum floor), and 14 per cent is living in the Kandi (lower hilly slopes) areas and the side valleys within 2,000 m above the sea level, while the remaining one per cent is sprinkled in the high latitudinal zones of the Himalayas, Zanskar, Ladakh and Karakoram Ranges.

Population enumeration was not done in the state of Jammu and Kashmir in 1991 owing to political instability. The projected population as reported by the Census of India in 1991 was about 77.19 million with a density of 76 persons per sq km. In 1996, the population of the state grew to about 89 millions which is likely to become 115.9 millions in 2011 A.D. as per the projected data published by the UNO Population Division 1998. Some basic information about the population of the state in 1981-91 have been given in Table 7.1.

It may also be observed from Table 7.1 that in 1991 the sex ratio was 923 females per 1,000 of males which was below the national average of 927. Nearly 24 per cent of the total population was urban and the remaining 76 per cent was rural largely dependent on agriculture, cattle-keeping and forestry. In 1991, the total urban population was 18.40 lakhs, out of which 6.09 lakhs was the slum population. The sex ratio declined from 925 in 1981 to 923 in 1991. The literacy rate was about 27 per cent in 1981 which rose to 46 per cent in 1991. The decadal growth rate of population between 1981-91 was about 29 per cent. The increase in population is a direct outcome of differential between the birth and the death rates as immigration of population in the state of Jammu and Kashmir is almost insignificant.

Table 7.1

Jammu and Kashmir: Demographic Attributes 1981-91

		1981	*1991 (Projected)*
1.	Total population	59,87,389	77,18,700
2.	Density of population per sq km	59	76
3.	Decadal percentage growth rate	29.69	28.92
4.	Urban population	21.05	23.83
5.	Sex-ratio	925	923
6.	Literacy rate	26.67	46.25

Source: Census of India, 1991.

Growth of Population

The size of population and its growth trend have a direct bearing on the economic development, social well-being, cultural values and political stability of a region. Population growth is, thus, pivotal to the regions demographic dynamism.[1] It is this attribute with which all other characteristics of population are intimately related. The change in population can be measured both in terms of absolute numbers and in terms of percentage.

In India, enumeration of population is being carried on since 1871, but reliable data on the population of Jammu and Kashmir is far from satisfactory. The state has had a very checkered demographic history. In general, prior to 1921, the population of the state grew at a slow pace. It was because the state experienced a number of famines and epidemics. After 1921, however, the population increased steadily. In 1951, just after independence, the population of the state was 32.5 million. In 1961, it grew to 35.6 million, which went upto 46.1 million in 1971. The last census in the state was held in 1981 which recorded a population of about 59.5 million (Fig. 7.1). The projected population for 1991 was 77.19 million, while the estimated population for 2001 and 2011 are 101.23 and 117.88 million respectively Table 7.2. The growth pattern of population between 1971 and 1981 has been plotted in Figure 7.2.

The growth of population in the state during the last five decades has been stupendous. This phenomenal increase in population may be attributed to the declining death rate and constant high birth rate. As a matter of fact, the medical service has been extended in the rural areas as a result of which the death rate has declined substantially. The district-wise population of the state has been given in Table 7.3 and plotted in Fig. 7.3.

Table 7.2

Jammu and Kashmir State: Growth of Population 1951-2011

Year	*Population in Million*	*Growth Rate*
1901	21.39	—
1951	32.5*	0.84
1961	35.6	9.54
1971	46.1	29.49
1981	59.5	29.06
1991*	77.19*	28.95*
2001*	101.23*	25.00
2011*	117.88*	23.95

Sources: 1. Registrar General, India, Population Projections for India and States, 1996-2016.
2. Report of the Technical Group on Population Projections, August, 1996.
3. Census of India.

Table 7.3

Jammu and Kashmir Population Distribution : District-wise-1981

	District	*Population*	*Percentage*	*Total*
1.	Doda	424,390	7.13	
2.	Jammu	928,638	15.60	
3.	Kathua	365,029	6.13	
4.	Poonch	222,490	3.74	
5.	Rajauri	298,830	5.01	
6.	Udhampur	452,244	5.60	
	Jammu Division	2,691,621	—	45.20
7.	Anantnag	654,945	11.00	
8.	Budgam	352,759	5.92	
9.	Barmulla	669,405	11.24	
10.	Kupwara	327,552	5.50	
11.	Pulwama	404,351	6.79	
12.	Srinagar	721,078	12.11	
	Kashmir Division	3,139,90	—	52.58
13.	Leh	67,733	1.14	
14.	Kargil	64,566	1.08	
	Ladakh Division	132,299	—	2.22
	Total Jammu and Kashmir State	5,954,010	100.00	100.00

Source: Census of India, 1981.

FIGURE 7.1

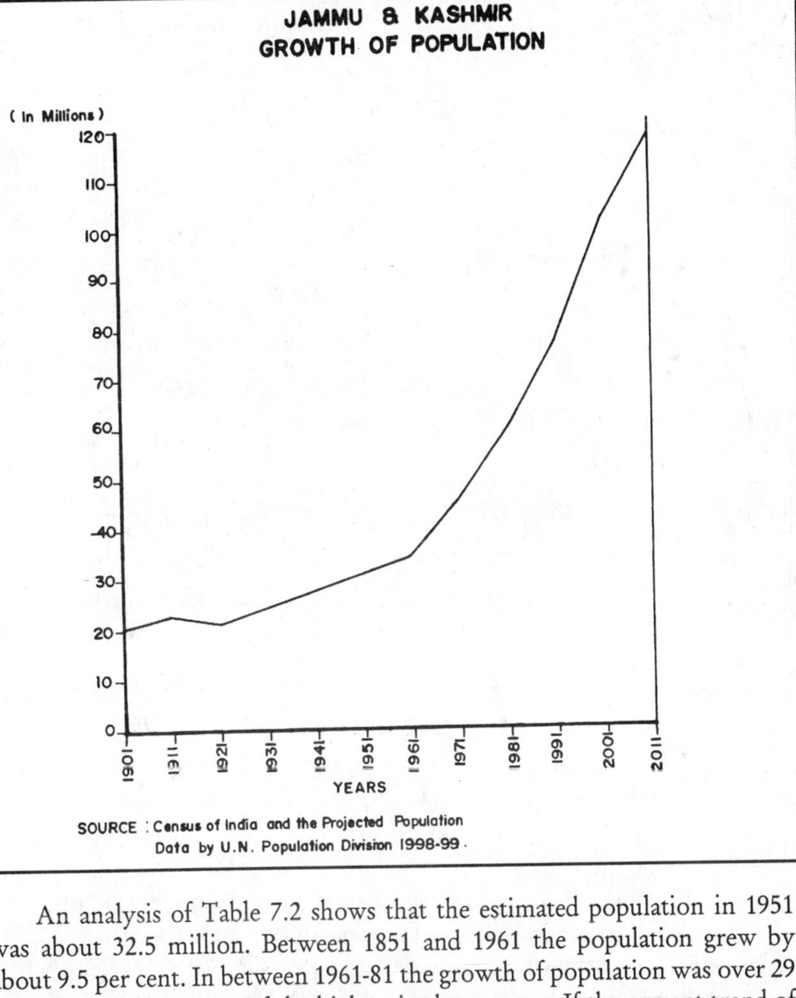

SOURCE : Census of India and the Projected Population Data by U.N. Population Division 1998-99.

An analysis of Table 7.2 shows that the estimated population in 1951 was about 32.5 million. Between 1851 and 1961 the population grew by about 9.5 per cent. In between 1961-81 the growth of population was over 29 per cent which was one of the highest in the country. If the present trend of population growth continues, there is a strong possibility that in 2021 A.D. the total population of the state would be over 125 million. Another notable feature of population has been the increase in the life expectancy. The male and female life expectancy in the state is about 57 and 60 years respectively. The life expectancy of males shows that the state has reached the take-off steps in the demographic evolution. The longevity in the state may go up if state are taken to increase literacy rate, particularly that of the females by improving levels of educational attainment and enhancing age at marriage.

FIGURE 7.2

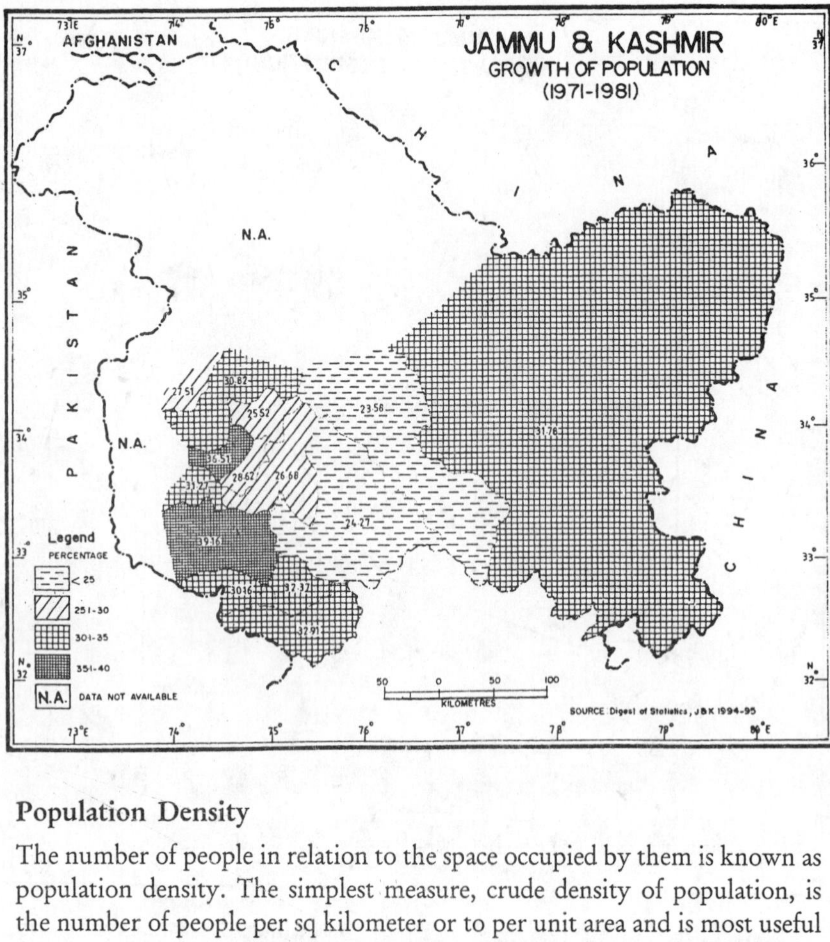

Population Density

The number of people in relation to the space occupied by them is known as population density. The simplest measure, crude density of population, is the number of people per sq kilometer or to per unit area and is most useful for small units such as districts and blocks rather than national states or continents, where internal environmental conditions vary markedly. More refined density measure may be calculated by relating numbers to cultivated or cultivable land or to other economic measures such as national income. Equally, for urban areas, measures such as persons per room or per house are useful refinements[2]. In brief, population density is a *land-men ratio*.

The density of population is a useful concept and a useful tool in the analysis of diversity of men's distribution in space. The geographers, by way of manipulating the numerator or denominator, have devised various types of densities having a varying degree of utility in different situations. The ob-

FIGURE 7.3

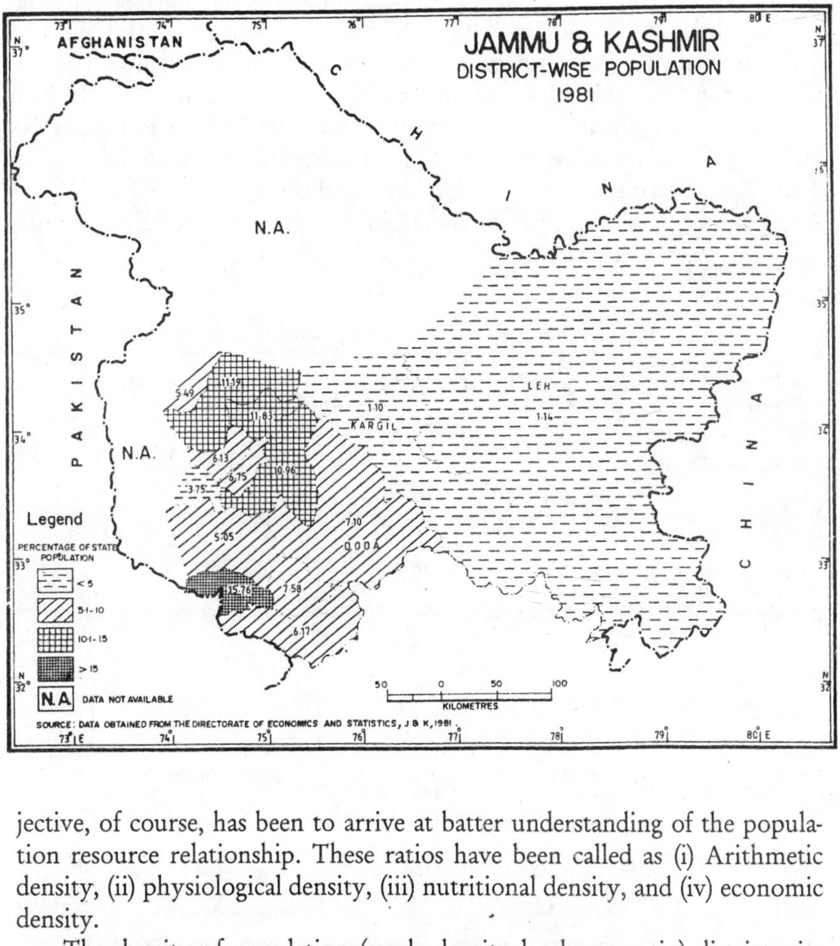

jective, of course, has been to arrive at batter understanding of the population resource relationship. These ratios have been called as (i) Arithmetic density, (ii) physiological density, (iii) nutritional density, and (iv) economic density.

The density of population (crude density land-men ratio) district-wise has been shown in (Fig.7.4 Table 7.4). An examination of this Table shows that the average density of the state in 1981 was 59 person per sq km which rose to about 76 persons per sq km in 1991. In comparison to the national average density of 276, the density of population in the state of Jammu and Kashmir may be said as sparse. Within the state, however, there are significant variations in the average density of population (Table 7.4). It may be observed from Fig.7.4 (Table 7.4) that there are four districts, namely, Srinagar, Jammu, Pulwama and Badgam in which the average density of population is much higher to that of the national average of 276. The highest density is in Srinagar District being over 400 persons per sq km followed by

Jammu District in which the density is about 390 persons per sq km (Fig. 7.4). In Anantnag, Baramulla Kathua, Kupwara, Poonch, Rajauri, Doda and Udhampur districts the average density of population is more than 100 persons per sq km. There are two districts, namely, Kargil and Leh (Ladakh Division) in which the density is only 36 and 2 persons per sq km respectively. The district of Leh is one of the most sparsely populated district of India. The difficult terrain, harsh climatic conditions, scanty rains, and poor accessibility are the main factors responsible for the sparse population in the hilly and mountainous districts of the Ladakh Division.

Table 7.4

Jammu and Kashmir: Density of Population 1981

Unit	*Density of Population per sq km*	*No. of Households per sq km*
All India	216	38.79
Jammu and Kashmir	59*	9.27
Srinagar	318	46.33
Jammu	305	52.63
Pulwama	289	41.76
Budgam	268	39.75
Anantnag	165	24.57
Baramulla	146	21.03
Kathua	139	23.19
Kupwara	138	21.47
Poonch	134	20.65
Rajauri	115	18.48
Udhampur	100	17.47
Doda	100	5.71
Kargil	36	0.88
Leh	2*	0.33

* Density has been worked out on comparable data.

Source: Digest of Statistics, Directorate of Economics and Statistics. Jammu and Kashmir. 1994.

The density of population per household has also been worked out and given in the third column of Table 7.4. The household density per sq km also reveals an identical pattern. For example, the districts of Srinagar, Jammu, Pulwama and Badgam have about 40 or more households per sq km while the state average is only about a households per sq km. The districts of Doda, Kargil and Leh have less than 9 households per sq km. The variations

FIGURE 7.4

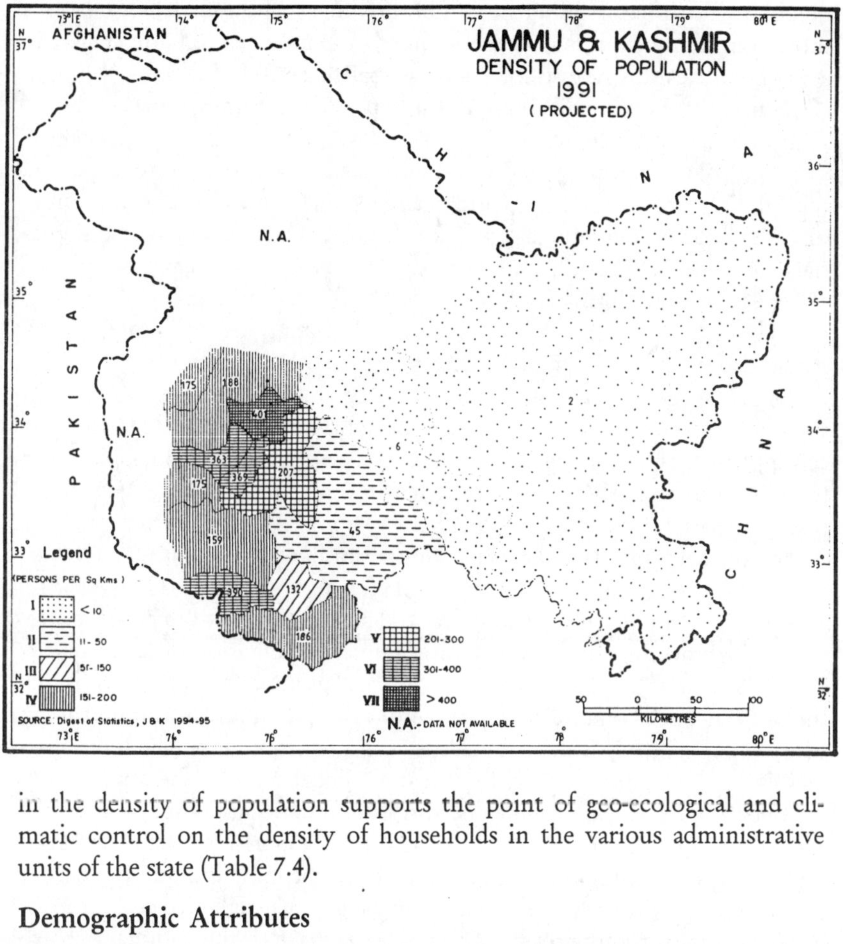

in the density of population supports the point of geo-ecological and climatic control on the density of households in the various administrative units of the state (Table 7.4).

Demographic Attributes

The size of population, its birth rate, death rate, life expectancy, age and sex composition, occupational structure, literacy rate, religious composition and migration pattern are the major attributes of population which determine the quality of human resource, social milieu and economic prosperity of the population of a region or nation state. The salient demographic characteristics of the Jammu and Kashmir state have been briefly given in the following paragraphs.

Sex Composition

The composition of a population according to sex is known as sex composition. The sex composition of a population helps in understanding the demographic processes of fertility, mortality and migration. The sex ratio is most commonly expressed as the number of females per 1000 of males. In some countries this ratio is expressed as males per 100 of females, while in countries lika Russia and Ukraine it is expressed in terms of percentage of male or female population.

The role of an individual in a family and society is deeply associated with the age and sex. The sex ratio is largely determined by a large number of genetic-socio-economic and political factors, which include the differentials of birth and death rates among the males and females and the sex-selective migration.

The sex ratio in the state of Jammu and Kashmir, according to the estimated enumeration of 1991 was 923 females per 1000 of males, as against the national average of 927 and the world average of 993. The low sex ratio in the state may be attributed to the high mortality rate among the females. Excepting the city of Srinagar, Jammu and a few affluent peoples in other towns, it has been observed that the females in general and the Muslim females in particular are the ignored lots. This biasness is more significant among the rural masses and among the transhumants like Gujjars and Bakarwals.[3]

The spatial pattern of sex ratio at the macro, meso and micro-levels reveals a high degree of variation. For example, the Division of Kashmir has only 878 females per 1000 of males, while in Jammu and Ladakh, Divisions the sex ratio is 925 and 879 respectively. The Division of Kashmir having over 95 per cent of Muslim population has a low sex ratio. The female babies, it seems, are not properly looked after. The families in the rural society are overworked as they have to perform the indoor and outdoor activities and nurturing of babies and children is their tiresome responsibility. Contrary to this, the Jammu Division has an overwhelming Hindu population. The literacy rate among the female Hindus is much higher to that of Muslims. Consequently, their sex ratio is 925, very lose to the national overage of 927. In brief, in Jammu and Kashmir state, the Muslim population is characterized by acute paucity of females. In fact, many of the Muslim boys are not getting married and in many cases the bridegrooms have to pay a substantial amount to the parents of brides. This social evil is the result of eschewed sex-ratio.

Age Structure

Age structure is an important characteristic of population which determines the workforce and dependency ratio. There are three basic determinants of age composition of population. These include natality, mortality and mobility. These determinants of age structure are interdependent and any change

FIGURE 7.5

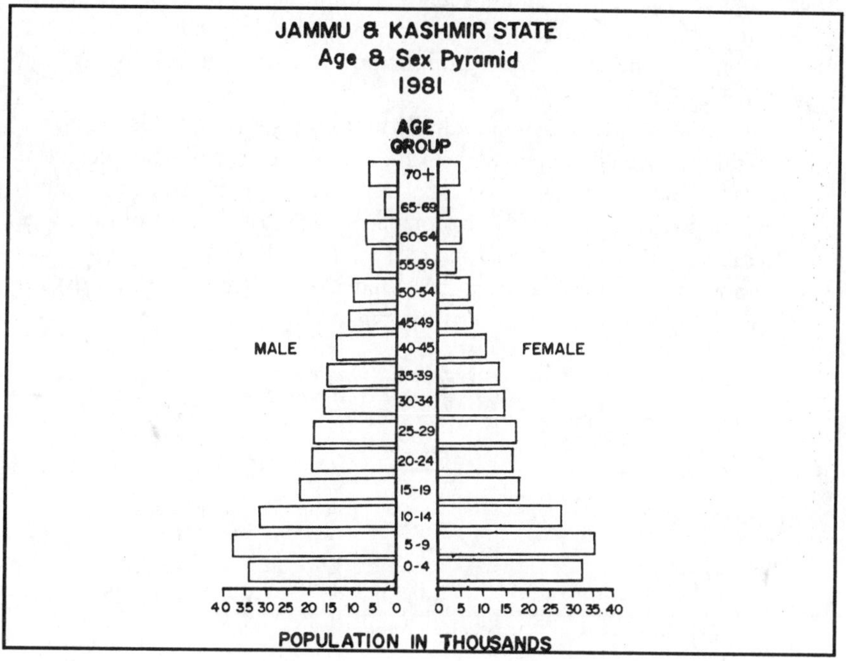

in one of them may influence the other two and it is through these variables that the socio-economic conditions influence the age structure.

Apart from fertility, mortality and migration, the age structure of population is also influenced by wars (which are highly made selective in their causality), catastrophe, natural calamities, and population policies. The impact of wars is most pronounced on the age structure as these take a heavy toll of young males. The elimination of males of a particular age group also affects the fertility which in turn influences the age structure. Similarly, the natural calamities, catastrophe, as and when they occur, leave their own impact upon the age composition unless the affected age group has lived its entire age. Certain population policies also closely influence the age structure. For example, the 'one child policy' of China is resulting in reducing the number of younger population. The reduction in the population base of 0-4 age group in China will adversely affect the availability workforce after a few years. On the contrary, the dependent senile population will increase. The malignance of old and retired people will put heavy burden on the social security system of the country.[4]

The population of the state has been categorized into various age groups. Generally, the population is categorized into three broad age groups:

(i) The young-0-14 years, (ii) the adult, 15-59 years, and (iii) the old 60 years and above. The social and economic implications of these age groups are quite serious and therefore, deserve a systematic consideration and examination. The male and female population in various age groups has been given in Fig. 7.5 (Table 7.5).

The age structure is cartographically plotted on 'Age pyramids'. Population pyramid is a diagrammatic representation of the age and sex structure of population. Generally, the vertical axis represents age groups and the horizontal axis indicates the numerical or percentage distribution by sex. The population is a reflection of past and current demographic trends. A population whose pyramid has a broad base and narrows quickly upwards, for example, is a young and highly fertile population.

Table 7.5

Jammu and Kashmir-Age Groups, 1981

Age Group	*Male*	*Female*
0-4	6.85	6.50
5-9	7.65	7.50
10-14	6.00	5.90
15-19	5.10	4.80
20-24	5.00	4.80
25-29	4.00	3.90
30-34	3.90	3.70
35-39	3.35	3.00
40-44	2.25	2.05
45-49	2.20	2.00
50-54	1.31	1.20
55-59	1.29	1.12
60-64	1.23	1.10
65+	1.20	1.10
Total	51.33	48.67

Source: Census of India, 1981.

The age pyramid of the state of Jammu and Kashmir given in Fig. 7.5 is typical of an underdeveloped society. It has a very broad base, and a tapering top. Nearly 42 per cent of total population of the state is below the age of 15 years. Nearly 12 per cent is above the age of 59 Years. Thus, the dependent population is 56 per cent and 44 per cent is in the active age group. Each worker is supporting about 1.3 persons, besides himself. The dependency ratio in the state in significantly high which is a characteristic of the developing societies. The younger population will enter shortly into the work-force category. The states' potential for creating job opportunities must increase to meet the increasing demand for employment as a consequence of increasing proportion of adult population. Looking at the existing unemployment in the state, the government should concentrate to create more jobs, especially in the agriculture, dairying, forestry sericulture, floriculture, bee-keeping, industries and tourism.

Literacy Rate

Literacy has been defined as the ability to read and write with understanding. Literacy reflects the socio-economic and cultural milieu of a nation, ethnic group or community. The concept of literacy however, varies from country to country. The population commission of the United Nations considers the ability to both read and write a simple message with understanding in any language as a sufficient basis of classifying a person as literate. The Indian Census has adopted this definition and now many of the countries too are shifting to this definition. A distinction can however, be made between the literate and the educated. All those persons who are classified as literate on the basis of their ability to both read and write are further sub-divided into a number of categories on the basis of their length of schooling.

The distributional pattern of literacy rate in the state has been plotted in Fig. 7.6. According to the Census of 1981 only 26.67 per cent of the total population of the state was literate as against 36 per cent (1981) for the country as a whole. So for as the sex-wise literacy is concerned about 36 per cent males and only 16 per cent females were literate in 1981 (Table 7.5). Despite several steps by the state government, the literacy rate of the state is much below the national average of 52.11 (Table 7.6).[5]

Table 7.6

Jammu and Kashmir State: Literacy Rate 1981*

	Literacy Rate	*Male Literacy*	*Female Literacy*
Jammu and Kashmir	26.67	36.29	15.88
All India Average	36.23	46.89	24.82

* The percentages have been calculated on total population inclusive of the population in the age group of 0-5 Years.

Source: Census of India, 1981.

The district-wise literacy rate has been given in Table 7.6 and its patterns have been plotted in Fig. 7.6.

FIGURE 7.6

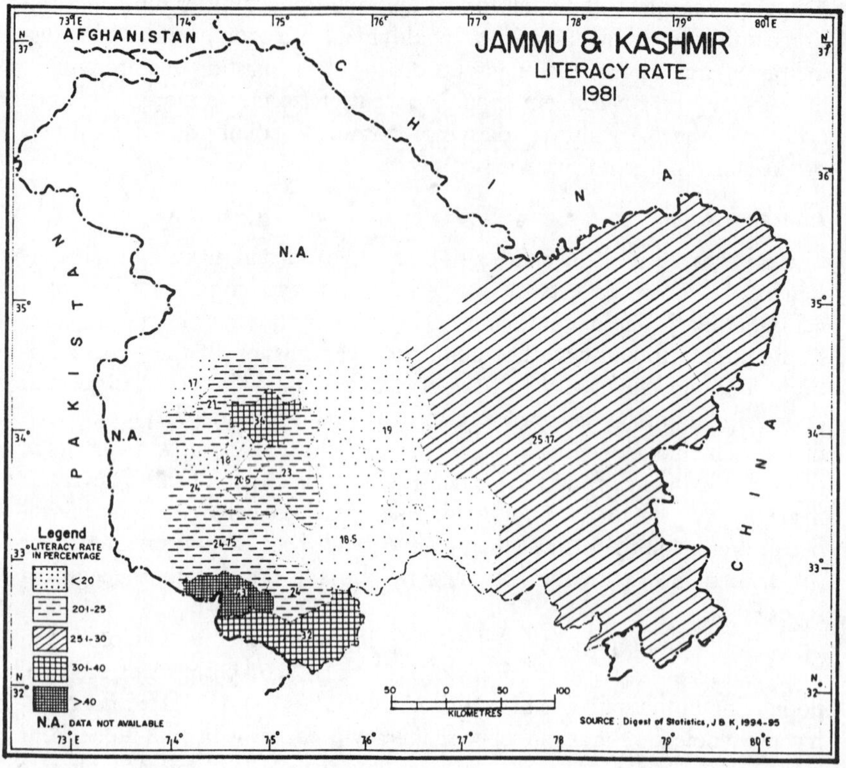

It may be observed from Table 7.7 that in 1981 in only four districts, namely: Srinagar (33.90%) Jammu (42.86%), Kathua (31.90%) and Leh (25.17%) the literacy rate was over 25 per cent, while in three districts, namely: Badgam (17.86%), Kargil (18.86%), and Doda (18.50%) less than 20 per cent of their population was literate. In the remaining districts, the literacy rate varied between 20 to 25 per cent (Table 7.7).

The disparity in the level of literacy in the rural and urban population of the state is quite conspicuous. About 26 per cent of the total population was literate in 1981, while the literacy rate in the rural masses was only 22 per cent. On the whole about 64 per cent males and 84 per cent females were illiterate in the state in 1981.

Rural literacy rate in all the three constituent units, i.e., Jammu, Kashmir and Ladakh Divisions is very low which shows the slow process of economic development. With the passage of time, the absolute number of illiterates is steadily increasing. Since economic development can not be achieved without literacy, education and technological skill, it is imperative to accelerate the process of literacy and education in the state. Serious attention is to be paid to educate the younger generation of the Gujjars and Bakarwals and those who are living in the far flung rural areas. The socially backward and economically deprived people deserve special attention for the eradicating illiteracy from the tradition bound social groups of the state.

Table 7.7

Jammu and Kashmir State: Literacy Rate-Districtwise, 1981

	District/State	*Persons*	*Literacy %*
1.	Anantnag	656351	22.93
2.	Pulwama	404078	20.47
3.	Srinagar	708328	33.90
4.	Budgam	367262	17.86
5.	Baramulla	670142	20.62
6.	Kupwara	328743	16.82
7.	Leh	68380	25.17
8.	Kargil	65991	18.86
9.	Jammu	943395	42.86
10.	Udhampur	453636	23.50
11.	Doda	425262	18.50
12.	Kathua	369123	31.90
13.	Rajauri	302500	24.73
14.	Poonch	224197	23.39
	Jammu and Kashmir State	5987389	26.67

Source: Census of India, 1981.

Religious Composition

Religion has been defined as the "feeling of absolute dependence". Religion is a symbol of group identity and a cultural rallying point. All societies have *value Systems*, i.e., common beliefs, understandings, expectations, and controls that unite their members, and set them off from other different cultural groups. Such a value system is termed as religion when it involves system of formal or informal worship and faith in the sacred and divine (God).

All the religions are concerned with the well-being of mankind on earth. All religions have left an imprint on the customs, traditions, fine arts and over all gamut of life of the society. The basic needs like food, clothing, shelter and occupations of the people are closely influenced by the religious values. Religion has been divisive as well as unifying force. The religious composition of a state, thus, has great social and economic relevance. The religious composition of the Jammu and Kashmir state has been given is Table 7.8.

Table 7.8

Jammu and Kashmir State: Religious Composition, 1981

Division	*Muslims*	*Hindus*	*Sikhs*	*Buddhists*	*Others*
Kashmir	94.00	4.42	1.20	0.05	0.33
Jammu	33.81	62.06	3.67	0.11	0.35
Ladakh	44.66	0.01	—	51.82	3.51
Total	65.84	30.41	1.06	1.26	0.19

Source: Census of India, 1981.

An examination of Table 7.8 shows that nearly 66 per cent of the total population of the state of Jammu and Kashmir consists of Muslims and 30.41 per cent of Hindus. Buddhists with 1.26 per cent constitute the third largest religious group followed by Sikhs (1.06%) in the state.

There are however, significant spatial variations in the concentration of different religious groups. In the Valley of Kashmir, Muslims have an overwhelming majority constituting 94 per cent of the population, and in Jammu Division about 62 per cent are Hindus, while in Ladakh Division 51.82 per cent are Buddhists. Thus, it may be said that Jammu is the Hindu dominated area, Kashmir has significantly large proportion of Muslims, while in Ladakh over 50 per cent of the population is Buddhist. The spatial variation in the population of different districts are quite significant. There are many districts in which either Muslim, Hindus or the Buddhists have overwhelming dominance. The district, wise distribution of population of the various religious groups has been given in Table 7.9.

An examination of Table 7.9 vividly shows that Muslims have an overwhelming majority in Kupwara (95.6%), Baramulla (96.5%), Pulwama (96%), Budagam (96%), Anantnag (95.6%) and Srinagar (90.7%). In Poonch (Jammu division) and Kargil (Ladakh Division) the percentage share of Muslims is 89 per cent and 78 per cent respectively. Rajauri and Doda are also the districts in which Muslims constitute over 58 per cent and 57 per cent of their total population. Muslims are insignificant in the districts of Jammu (4%), Kathua (7%) and Leh (15%).

Contrary to this the proportion of Hindus is significantly high in the district of Kathua (91%), Jammu (87%) and Udhampur (72%).The Doda and Rajauri Districts have about 42 and 39 per cent of Hindu population, while

Table 7.9
Jammu and Kashmir: Religious Composition, 1981

District	Total Population	Percentage			
		Muslims	Hindus	Sikhs	Others
Anantnag	656351	95.63	3.76	0.59	0.02
Budgam	367262	95.94	2.63	1.40	0.03
Baramulla	670142	96.50	2.02	1.46	0.02
Kupwara	326743	98.12	2.03	0.44.	
Pulwama	404078	95.87	1.54	1.62	0.07
Srinagar	708328	90.65	8.38	0.89	0.08
Leh	68380	15.31	2.99	0.27	81.43
Kargil	65992	77.90	2.26	0.22	19.22
Doda	425262	57.30	42.22	0.22	0.26
Jammu	943395	4.27	87.37	7.64	0.72
Kathua	369123	6.96	91.16	1.65	0.23
Punch	224197	89.06	7.29	3.62	0.03
Rajauri	302500	58.28	39.24	2.42	0.06
Udhampur	453636	26.23	72.20	1.27	0.03
Total	5987289	64.19	32.24	2.23	1.34

Source: Census of India 1981.

their percentage share in all the districts of Kashmir Division is around 2 or less than 2 per cent except Srinagar District in which Hindus constitute about 8 per cent of the total population. Their proportion in Leh and Kargil districts is about 8 and 3 per cent respectively. In the Leh District of Ladakh the main ethnic group is Buddhists. The percentage share of Buddhists population in Leh and Kargil districts is 81 and 20 per cent respectively.

Workforce

Any group of population consists of workers and non-workers. The workers are referred as manpower. The manpower consists of only those persons who could participate in economically gainful activities in the event of need. Different countries further classify manpower into: (i) economically active population, and (ii) economically inactive population. The economically active population is engaged in the production of goods and services. It consists of both males and females. The inactive population is that which is engaged in activities like household duties in their own house or at the place of their relatives, retired personal, inmates of institutions, students and those living on royalties, rents, dividends, and pensions.

The population of a region is divisible into employed (workers), and unemployed (non-workers). The employed active population consists of all those persons who are at work during the reference period of the census including those who are temporarily absent from work due to illness, and industrial dispute. By comparison, the unemployed, economically inactive population consists of those who were not at work during the reference period of the census but are seeking some work.

The Indian census divides the entire population into two categories of workers and non-workers. According to the census of India, any person whose main activity was participation in any economically productive work either by his physical or by his mental activity was classified as worker. Thus, work involved not only actual work but also effective supervision and direction. The workers have been sub-divided into: (i) main workers, and (ii) marginal workers. The main workers are those who worked in some economic activity for more than six months (183 days). Marginal workers on the other hand, are those who have worked any time at all in the year preceding the enumeration but have not worked for six months in the year.

The workforce in the state of Jammu and Kashmir at the district level as per the census of 1981 has been given in Table 7.10 and plotted in Fig. 7.7.

Table 7.10

Jammu and Kashmir State: Workers and Non-Workers, 1981

	District/state	*Population*	*Main Workers*	*Marginal Workers*	*Non Workers*
1.	Anantnag	656351	31.68	10.77	57.56
2.	Pulwama	404078	29.04	21.66	49.30
3.	Srinagar	708328	30.12	4.76	65.10
4.	Budgam	367262	32.81	15.92	51.27
5.	Baramulla	670142	31.43	14.02	54.55
6.	Kupwara	328743	30.37	18.49	51.14
7.	Leh	68380	43.55	6.84	49.61
8.	Kargil	65992	45.33	9.61	45.06
9.	Jammu	943395	26.66	12.77	60.57
10.	Udhampur	453636	31.62	22.96	45.42
11.	Doda	425262	33.48	9.88	56.64
12.	Kathua	369123	28.82	14.58	56.60
13.	Rajauri	302500	27.59	17.39	55.46
14.	Poonch	224197	27.72	18.82	53.46
	State	59,87,389	30.37	13.88	55.75

Source: Computed from the Census of India, 1981.

FIGURE 7.7

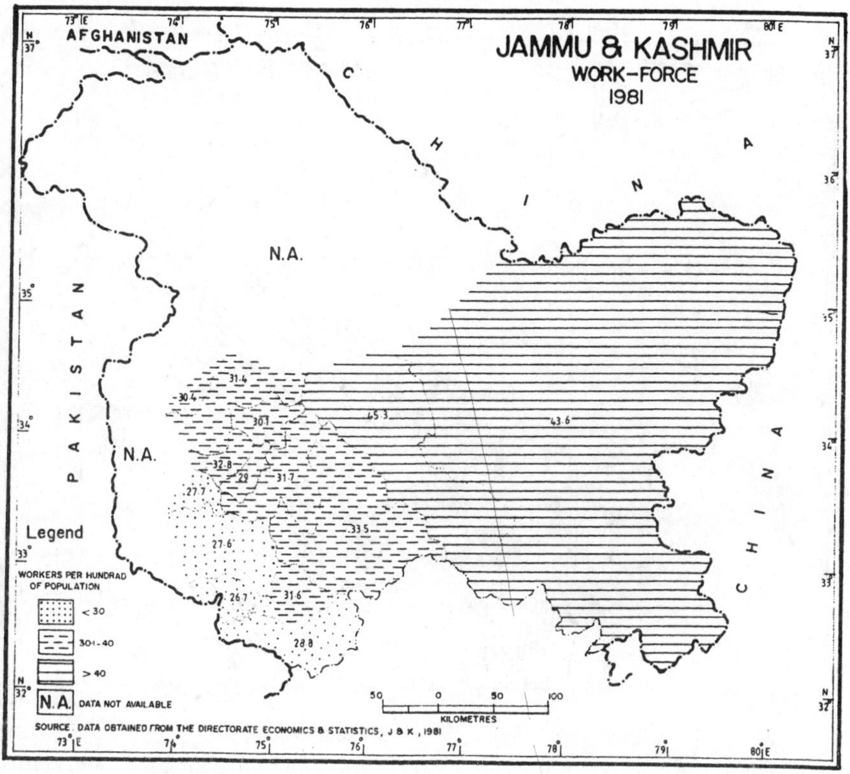

The workforce of a region/state is closely controlled by a large number of demographic factors. The size of population, the birth rate, the death rate, the longevity, migration pattern, literacy, education, status of women is the society, age at the time of marriage, general health of the people, the occupation of the people, their dietary habits, per capita income attitude towards life and standard of living influence the availability of workforce in a society.

The state of Jammu and Kashmir is characterized by a low participation rate. It may be observed from Table 7.10 that only about 30 per cent was the workforce (main workers) in 1981 and about 14 per cent were marginal workers, who worked for less than 183 days in a year. The non-workers (dependents) constituted about 56 per cent of the total population.

The low proportion of the workforce and the high dependency ratio in the state of Jammu and Kashmir was mainly because of the high birth rate and orthodoxy in the society which restricts the females to participate in the

FIGURE 7.8

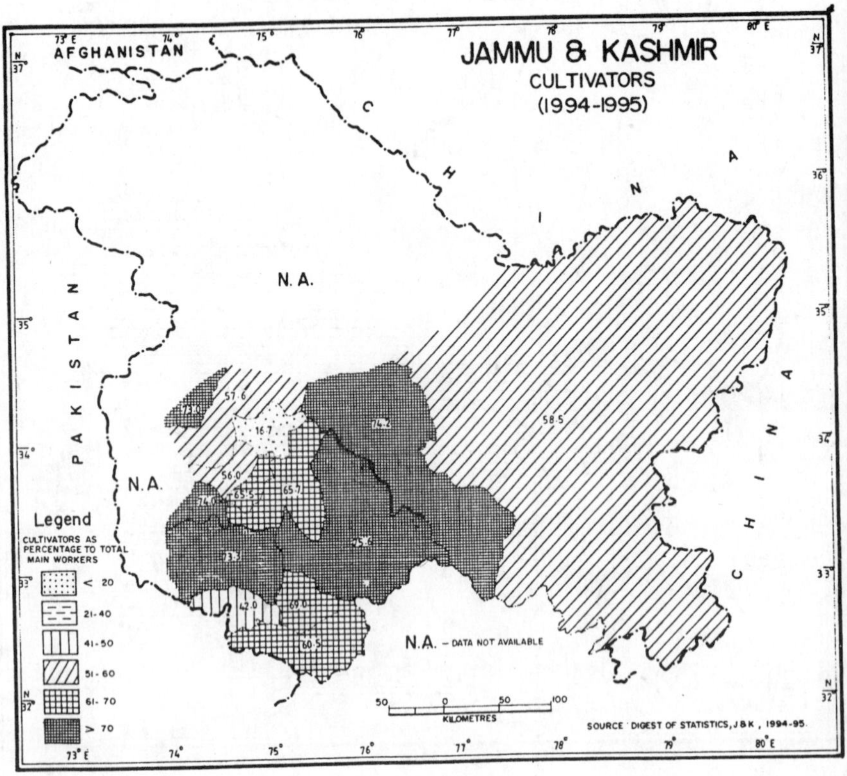

productive economic activities. It appears that in the tradition bound society of Kashmir the rural females are not getting proper medical care and education. Their status in the society is low and the high birth rate limits job opportunities for them. Consequently, the responsibility of bread-earning is mainly on the shoulders of male workers.

The district-wise workforce as given in Table 7.10 shows that in five districts (Pulwama, Jammu, Katnua, Rajauri, Poonch, the workforce is less than 30 per cent. In other words, the dependency ratio in these districts is significantly high, being more than 2.25 persons per worker. Interestingly, enough the districts of Kargil and Leh have the highest workforce, being over 45 and 43 per cent of their total population respectively. In the remaining districts of Jammu and Kashmir Divisions the workforce varies between 30 to 33 per cent (Table 7.10).

The main workers if classified on the basis of age and sex show that child labour is much prevalent in the state. The age and sex-wise break-up of workforce has been given in Table 7.11.

An analysis of Table 7.11 shows that about 6 per cent of the workforce is below the age of 15 years which means child labour is quite prevalent in the state. The economically poor people are not in a position to send their children to school. They engage them in either their agricultural activities or send them to cottage industries, carpet making, etc., to fetch some money for the sustenance of the family. The child labour may be observed all over the state, especially in the urban areas like Srinagar, Jammu, Baramulla and Anantnag. Child labour is not confined to boys, as even the girls below 15 years of age to the extent of 1.27 per cent of the total workforce are engaged in economically gainful employment (Table 7.11).

Table 7.11

Jammu and Kashmir State: Workers by Age and Sex, 1981.

	Workers		
	Total	*Male*	*Female*
0-14	5.99	4.72	1.27
15-19	11.06	9.70	1.36
20-24	13.20	11.84	1.36
25-29	13.29	11.97	1.32
30-34	11.36	10.31	1.05
35-39	10.86	9.98	0.88
40-45	8.52	7.74	0.78
45-49	8.27	7.62	0.65
50-54	5.55	5.03	0.52
55-59	4.50	4.13	0.37
60+	7.40	7.11	0.29
Total	100.00	88.41	11.59

Source: Census of India 1998.

It may also be observed from Table 7.11 that about 60 per cent of the total workforce come from the age group between 15 to 39 years. This is the economically most productive and sexually most reproductive age group. Crossing the age of 40 years, there is a gradual decline in the percentage of workforce. There are, however, about 7 per cent of the main workers who are more than 60 years of age. It is in the agriculture, medicine, legal profession and business in which even beyond 60 years of age people keep themselves busy into some economically gainful employment. In general, the male and female proportion in the workforce is about 88 and 12 per cent respectively (Table 7.11).

Occupational Structure

Occupational structure means the division of working population in different occupations and professions. In each society, people engage themselves in different productive economic activities. The occupation of an individual refers to his trade, profession, and type of work. The occupational structure of a society is closely influenced by a number of geoclimatic, socio-economic, political factors and human aspirations. The nature and variety of physical resource base, like arable land, forest water and minerals also determine the occupational structure of the people of a region.

The occupational structure is often been classified into: (i) primary, (ii) secondary and (iii) tertiary categories. The number of occupations are however, numerous. The major occupations of the workforce of the state have been classified on the basis of the National Industrial Classification and are given in Table 7.12.

Table 7.12

Jammu and Kashmir State: Major Occupations, 1984

S.No.	*Occupation*	*Percentage*
1.	Cultivators	65.0
2.	Agricultural labourers	3.2
3.	Livestock forestry, hunting, plantation, orchards and allied activities	3.8
4.	Mining and quarrying	0.1
5.	Manufacturing, processing, service and repairs (household and others)	6.8
6.	Construction	2.4
7.	Trade and commerce	4.5
8.	Transport, storage and communication	2.4
9.	Other services	11.8
	Total	100.00

Source: Digest of statistics, Jammu and Kashmir, 1984.

It may be observed from Table 7.12 that about 68 per cent of the total workforce of the state was engaged in agriculture, out of which 65 per cent were cultivators and 3.2 per cent agricultural workers. About 4 per cent of the workers were engaged in other primary activities like livestock, forestry, hunting, plantation, orchards keeping, mining and quarrying. Manufacturing, processing and repairs provided jobs to about 7 per cent of the workers,

FIGURE 7.9

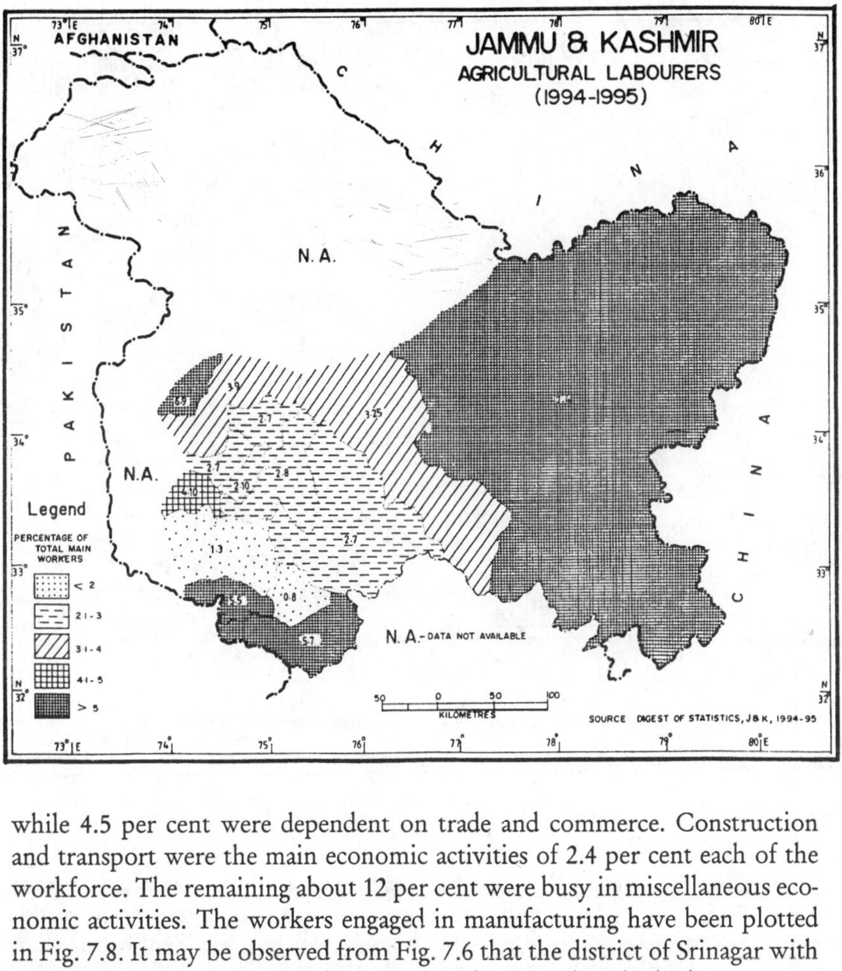

while 4.5 per cent were dependent on trade and commerce. Construction and transport were the main economic activities of 2.4 per cent each of the workforce. The remaining about 12 per cent were busy in miscellaneous economic activities. The workers engaged in manufacturing have been plotted in Fig. 7.8. It may be observed from Fig. 7.6 that the district of Srinagar with over 30 per cent of the workforce in manufacturing has the highest concentration of industrial workers. It is followed by Badgam District in which over 22 per cent are engaged in manufacturing. In general the proportion of workers engaged in manufacturing is higher in the districts of Kashmir Division as compared to those of the Jammu and Ladakh Divisions (Fig. 7.10).

A close examination of the workers and their main economic activities show that the state of Jammu and Kashmir is in the total workforce is dependent of primary activities. The service sector is providing jobs to over 21 per cent of the workforce, while the remaining about 7 per cent is dependent on the secondary sector, i.e., industries.

FIGURE 7.10

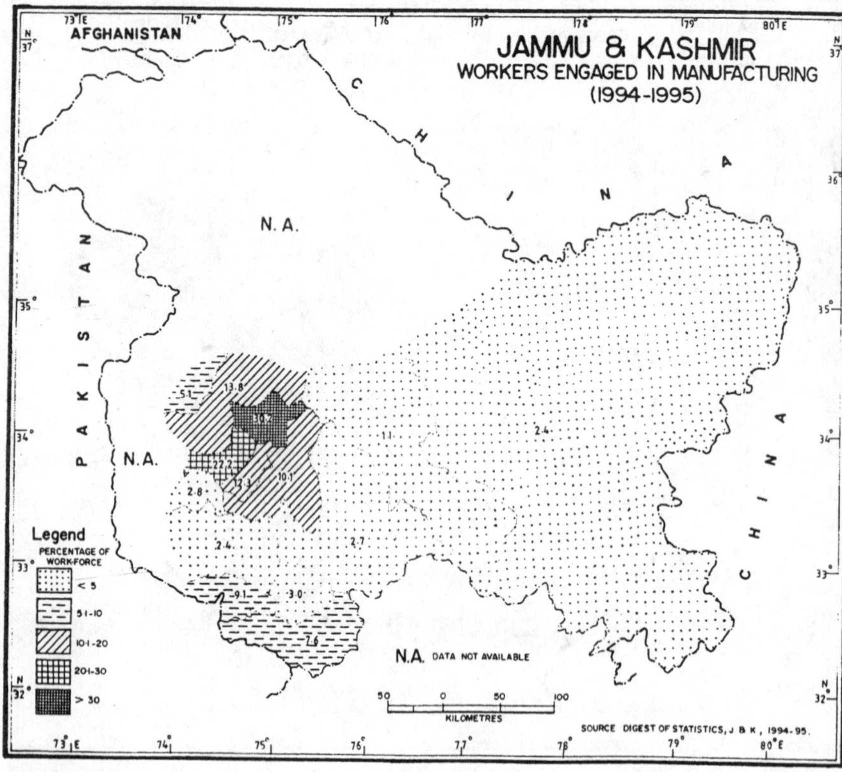

Cultivators

As stated at the outset, the state of Jammu and Kashmir is essentially agrarian of which about 75 per cent of the workforce is directly or indirectly dependent on primary activities. The spatial distribution of cultivators has been shown in Fig. 7.8.

An examination of Fig. 7.8 shows that in five districts, namely: Doda (75.6%), Poonch (74%), Rajauri (73%) and Kargil (74%), over 70 per cent of the workforce is engaged in agriculture. In Kupwara and Baramulla, the cultivators constitute 57.6 and 56 per cent of the workforce respectively. Jammu District with 40 per cent and Srinagar District with 16.7 per cent have low proportion of cultivators in the workforce as they are more urbanized.

Agricultural Labourers

The district-wise distribution of agricultural workers has been plotted in

Fig. 7.9. It may be observed from Fig. 7.9 that the districts of Kathua and Jammu have a relatively high proportion of agricultural workers in Jammu Division, being 5.7 and 5.5 per cent of the total workforce respectively. In the Kupwara District of Kashmir Division and Leh District of Ladakh Division, the agricultural workers constitute about 6.9 per cent and 7.4 per cent of the total workforce (Fig. 7.9). In rest of the districts this proportion is less than 4 per cent. Interestingly enough, the districts of Udhampur (0.8%) and Rajauri (1.3%) have the lowest percentage of the agricultural workers.

In brief, the state of Jammu and Kashmir is in the initial stage of economic development and the workforce is largely engaged in the primary sector (agriculture, cattle-keeping, dairying, forestry, fishery sericulture, bee-keeping, floriculture, etc.). There is an urgent need to develop industries so that the pressure of population from the primary sector may be shifted to the secondary and tertiary sectors. The development of industrial sector will stimulate the economy by providing more jobs to the rural youths who are increasingly becoming unemployed owing to the small and marginal size of holdings. The industrial and economic development of the state, it is hoped, will ultimately lead to political stability. It is, therefore, imperative to formulate and implement a sound industrial policy. The establishment of agro-based, forest-based industries and tourism in the different parts of the state will go a long way in giving the state a sound industrial base on which the economy of the region may be developed. The rapid economic development will also help in overcoming many of the geo-political problems. It is a well-know axiom that economic development leads to political stability and political stability is the pre-requisite for economic development. For all This, dynamic leadership, participation of the people and entrepreneurship are required.

Notes

1. R.C. Chandna, 1992, *A Geography of Population*, Kalyani Pub. Ludhiana.
2. R.J. Johnston et. al., 1985, *A Dictionary of Human Geography*, Blackwell, Oxford.
3. Husain, M., 1986, *Geography of Jammu and Kashmir State*, Rajesh Pub. New Delhi 2.
4. *Census of India*, 1991.
5. UNFPA, 1997, *Towards Population and Development Goals,* Oxford University Press, New Delhi.

8

Agriculture

The practice of cultivating the soil in order to produce crops and domestication of animals and pastoral farming are known as agriculture. The agricultural processes of a region are directly controlled by the prevailing physical environmental condition (temperature, precipitation, terrain, soil etc.) and the socio-cultural milieu (land tenancy, size of holding, technology, workforce, family requirements, irrigation, power, roads, marketing, aspirations of the growers, etc.). The present chapter gives a concise account of the general landuse, cropping pattern, crops concentration, crops combinations and the agricultural operations being carried out in the different agro-climatic regions of the Jammu and Kashmir state.

Jammu and Kashmir is essentially a mountainous state in which only about 30 per cent of the reporting area is under cultivation. Agriculture is the mainstay of the people as it provides employment, directly or indirectly to about 70 per cent of the workforce. It contributes about 65 per cent of the state revenue which explains the overdependence of the state on agriculture. Land is, however, limited and therefore, its judicious utilization is necessary to meet the growing need of the tremendously increasing population and for the sustainability of soils, ecosystems and environment.

Land Utilization

Land utilization statistics provide the area figures, showing the distribution of the total geographical area/reporting area of a region/country into various uses. Although detailed statistics of landuse for the country, showing the area of land put to different uses, are continuously available since 1884, but in the case of Jammu and Kashmir there are several data gaps.

In India, in general, the land has been classified under different categories. In 1890-91, a five-fold classification of the total geographical area of the country was made and the land was put under the categories of: (i) forest, (ii) area not available for cultivation, (iii) current fallow, (iv) net area sown. In 1949-50, the land classification was, however, revised. The revised classification has been accepted for use by all the states of India, since 1950-51. The new classification has been introduced by all the states except West Bengal and Manipur where agricultural data is still recorded on the basis of old classification. The existing classification of landuse in the country is as follows:

1. Forest
2. Land not available for cultivation:
   (a) land put to non-agricultural uses, and
   (b) barren and uncultivated land.
3. Other uncultivated land excluding fallow lands:
   (a) permanent pastures and other grazing lands,
   (b) miscellaneous tree-crops and groves not included in the net area sown, and
   (c) cultivable waste.
4. Fallow land:
   (a) fallow land, other than current fallow, and
   (b) current fallow.
5. Net area sown.

General Landuse

As given in the first chapter, the total geographical area of the state is 2.23 lakh sq km including those parts which are under the occupation of Pakistan and China. About 92 per cent of the geographical area of the state consists of high mountains rugged topography and only 5 per cent is available for cultivation. The general picture of landuse and the proportion of area under different categories have been given in Fig. 8.1 (Table 8.1).

Table 8.1

Jammu and Kashmir-General Landuse, 1995-96

	Use of Land	*Total Reporting Area in (000 Hect.)*	*Percentage*
1.	Forest	658	27.24
2.	Net area sown	730	30.22
3.	Land put to non-agricultural uses	291	12.04
4.	Barren land	293	12.13
5.	Permanent pastures and other grazing grounds	125	5.17
6.	Land under miscellaneous trees and other groves	72	2.98
7.	Cultivable waste	141	5.84
8.	Fallow other than current fallow	7	0.29
9.	Current fallow	99	4.00
	Total	2416.00	100.00

Source: Digest of Statistics, Jammu and Kashmir, 1994-1995.

FIGURE 8.1

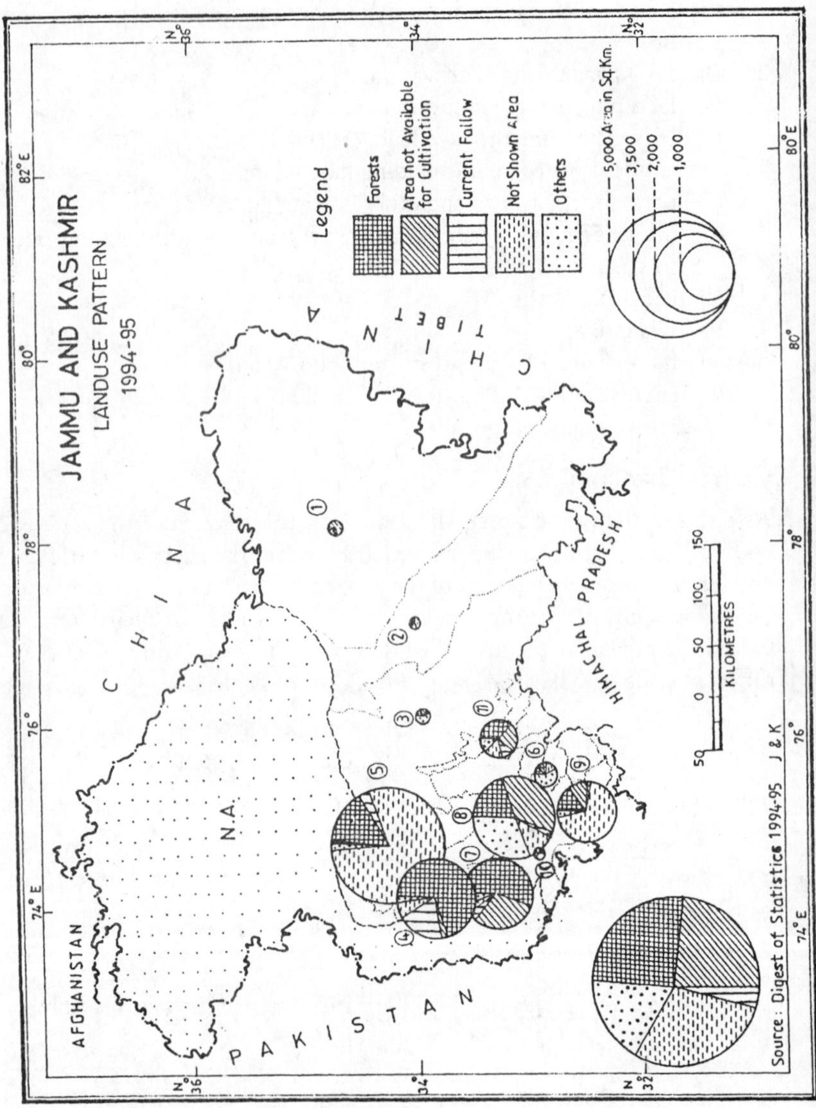

Source : Digest of Statistics 1994-95 J & K

Table 8.2

Jammu and Kashmir Land Utilization 1994-95. Area (000 ha)

Division	Total reporting area	Area under forest	Land not available for cultivation	Other uncultivated land excluding fallow	Fallow land old	Fallow land current	Net area sown	Total
Kashmir	557	6 (1.07)	93 (16.70)	87 (15.62)	2 (0.36)	23 (4.13)	346 (62.12)	100.00
Jammu	1794	652 (36.35)	457 (25.47)	238 (13.27)	4 (0.22)	73 (4.07)	370 (20.62)	100.00
Ladakh	65	—	33 (50.77)	12 (18.46)	2	1 (1.54)	19 (29.23)	100.00
Total	2416	658 (27.24)	583 (24.13)	337 (13.94)	6 (0.25)	97 (4.01)	735 (30.43)	100.00

Figures in parentheses show the percentages.
Source: Digest of Statistics, Jammu and Kashmir, 1994.

It will be seen from Table 8.2 that only 27.24 per cent of the total reporting area of the state is under forest which is below 33 per cent, recommended by the government under the National Forest Policy 1988. About 30 per cent is the net area sown. Permanent pastures as and other grazing grounds occupy about 5 per cent of the reporting area, while 12.13 per cent is barren land. Being, hilly, mountainous and snow covered, it is only the gentle slopes (below 15°) which may be developed as orchards and pastures after heavy investment. The proportion of old fallow and current fallow is 0.29 and 4.0 per cent respectively. About 12 per cent of the total reporting area is put to non-agricultural uses, e.g., settlement, roads, cemetry, *guls* (canals) and water bodies.

The net cultivated area has been plotted in Fig. 8.2. It may be seen from Fig. 8.2 that in the districts of Kashmir Division about 60 per cent or above of the total reporting area is under cultivation except in Ganderbal and Srinagar in which it is 47 and 49 per cent respectively. In Jammu Division the net cropped area varies between 14 per cent in Doda to 33 per cent Jammu District. In Ladakh, the net area sown is only about two per cent of the reporting area.

Cropping Patterns

Cropping pattern means the proportion of area under different crops at a point of time. The crop statistics published by the government are used to

FIGURE 8.2

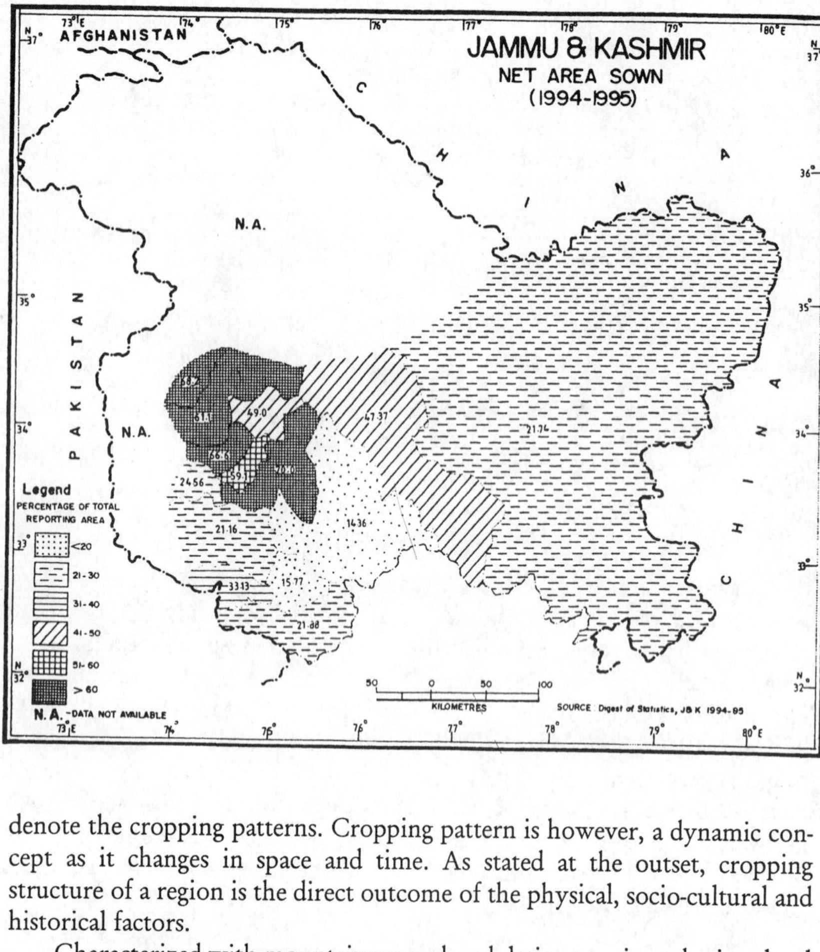

denote the cropping patterns. Cropping pattern is however, a dynamic concept as it changes in space and time. As stated at the outset, cropping structure of a region is the direct outcome of the physical, socio-cultural and historical factors.

Characterized with mountainous and undulating terrain and micro-level variations in temperature, precipitation and soils, the state of Jammu and Kashmir has a high degree of variation in its cropping patterns, crop combination and crop diversification. In general, the Jammu plain has a high concentration of wheat, rice, maize, pulses, fodder and oilseeds, while the Valley of Kashmir is well known for its paddy, maize, orchards (apples, almond, walnut, peach, cherry, etc.) and saffron cultivation. In Ladakh, barley, wheat, maize, vegetables, barseem and fodder are the main crops. The area and percentage of the total cropped area in the state have been given in Table 8.3.

Table 8.3

Jammu and Kashmir: Area under Different Crops 1995-96

Crop	*Area in thousand hectares*	*Percentage*
Rice	273.00	28.58
Bajra (Bulrush millet)	13.30	1.39
Maize	305.00	31.93
Wheat	250.00	26.17
Barley	7.68	0.80
Pulses	34.25	3.58
Sugarcane	0.32	0.03
Fruits and Vegetables	61.47	6.44
Condiments and spices	1.85	0.19
Other food and non food crops	7.24	0.76
Total	955.11	100.00

Source: Digest of Statistics, Govt. of Jammu and Kashmir, 1994.

An examination of Table 8.3 shows that maize is the first ranking crop in the state, occupying about one-third of the total cropped area. Rice with 28.58 per cent of the gross cropped area is the second most important staple crop, followed by wheat to which over 26 per cent of the total cropped area is devoted. It is interesting to note that rice and maize are the two rival crops for the first ranking, in the different parts of the state. Depending on the rainfall conditions, the hectarages under rice and maize fluctuate substantially. Consequently, in some of the years maize ranks first, while in others, rice occupies the maximum hectarage of the total cropped area. Out of these two staple food crops, rice has high concentration in the Kashmir Valley and wheat in the Jammu Plain. Wheat is mainly grown in the district of Kathua and Jammu, adjacent to the province of Punjab. Pulses, oilseeds, spices, fodder and vegetables are the other crops grown in the state (Table 8.3).

Maize (Zea-mays): *Maki*

In terms of hectarage, maize was the first ranking crop in Jammu and Kashmir in 1994-95. Nearly one-third of the total cropped area was devoted to its cultivation (Fig. 8.3, Table 8.3). It is the staple food of Gujjars and Bakarwals, living in the *Kandi* and hilly areas. Moreover, the grains form an important cattle food, being fed to farm cattle and horses. The different parts of the plant and the grain are put to a number of industrial uses. The silk threads of maize are used as a filter; husks for making of mattresses; cobs for the making of corn pipes. The stalks and cobs are made use of for the prepa-

FIGURE 8.3

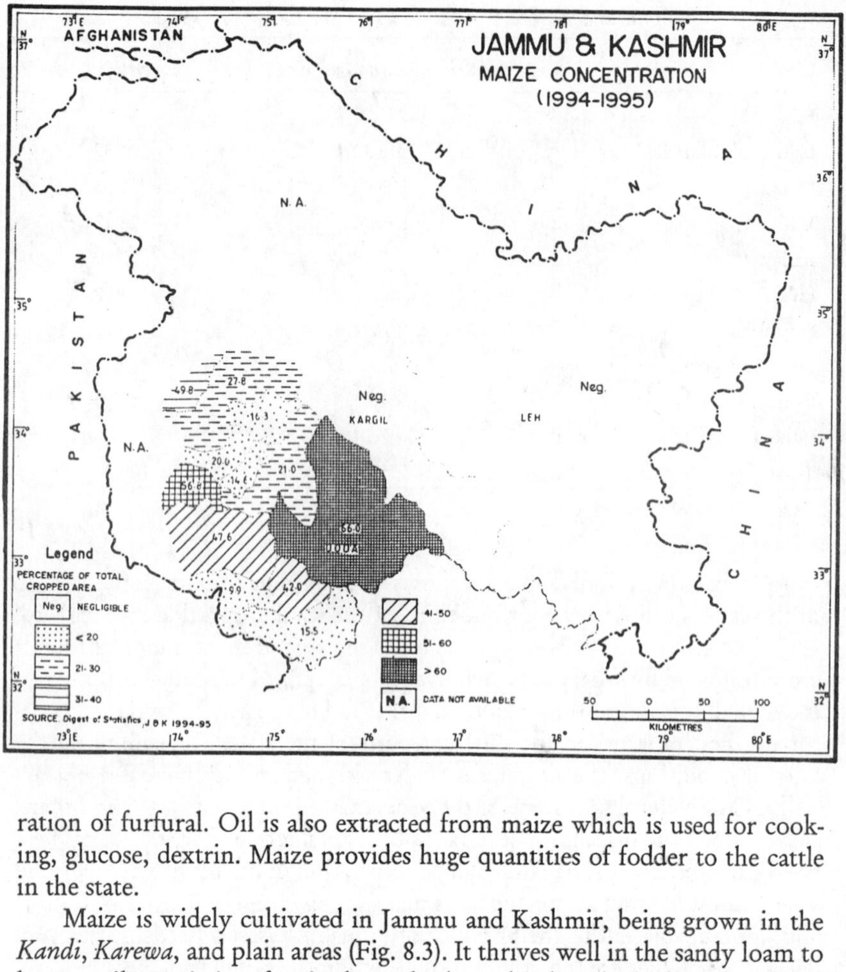

ration of furfural. Oil is also extracted from maize which is used for cooking, glucose, dextrin. Maize provides huge quantities of fodder to the cattle in the state.

Maize is widely cultivated in Jammu and Kashmir, being grown in the *Kandi*, *Karewa*, and plain areas (Fig. 8.3). It thrives well in the sandy loam to loamy soils. Varieties of maize have also been developed which perform well in the colder hilly and mountainous areas. It can be grown in all such regions where the summer is long enough to permit its cultivation, and frost does not set in too early. It needs about 30°C temperature at the time of germination, growth and development, and over 20°C at the time of ripening.

Rich, well-drained and light soils or Loamy in character are well suited for the maize crop in preference to heavy soils of clayey or clayey loam. The coarse soils on the hill slopes are also equally conducive for its cultivation. The crop requires adequate manuring even on soils which may be considered naturally fertile. In fact, maize is a soil exhaustive crop.

Maize requires a well-tilled and highly manured soil. The preparation of soil is carried out in time to suit the sowing season which is April to May in the Jammu Plain and May to June in the Valley of Kashmir and the Kandi and hilly areas of the state. The field is ploughed from two to three times or dug deeps with spade. Clods are broken by the female members of the cultivators' family. Before sowing, weeds and stubbles are removed and burnt. The field is ploughed two to three times to obtain a fine title. Cattle manure is also substantially applied to enhance the soil fertility.

The maize plant is subject to many of the insects and pests. The crop also suffers from insects attacks on the cobs. Now many plant protection chemicals have been developed to reduce the danger to the crop arising out of pests and disease.

Maize is grown in almost all the districts of the state except Leh and Kargil (Fig. 8.3). Its cultivation is largely confined to the *Kandi* areas and hilly tracts. The main concentration of maize is found in the Doda district in which it occupies about 82 per cent of the total cropped area. Rajauri, Poonch and Udhampur are the other districts in which its concentration is significantly high (Fig. 8.3).

Rice (unhusked or paddy) *Dhan* (Oryza-sativa)

Rice is the staple food in Kashmir, while its cultivation is widely carried on in Jammu Plain and in isolated tracts in the hilly areas, the side-valleys. This crop has the distinction of being the most extensively cultivated in Jammu and Kashmir state (Fig. 8.4).

Rice has an extensive range of cultivation. The range of altitude may be said to extend from 5 m to 1,800 m. Temperature, however, sets a limit and in the higher altitudes like, Kashmir Valley, Bhadarwah (Jammu Division), Rupsu, Nubra and Suru Valley (Ladakh Division). Rice can be grown only in the warm months of summer (July-Sept.). In the higher altitudes of mountains, where the temperatures are generally mild even during the summer season, the growing period is more prolonged and the ripening and harvesting get delayed. In fact, the short duration verities like *Pusa 103,* which mature in about 60 days in Haryana and Punjab, take about four months in the hilly and mountainous areas of the Jammu and Kashmir state.

Rice can be grown over a wide range of rainfall. In the areas of deficient rainfall, it is successfully grown with the help of irrigation. It requires about 100 to 125 cm of rainfall, well distributed over the growth period. The average temperature should be over 20°C. Frost, mist, fog, and low temperatures are injurious to its cultivation.

The soils on which rice can be grown are so varied that it can be said that there is hardly any type of soil in which it can not be grown, including

FIGURE 8.4

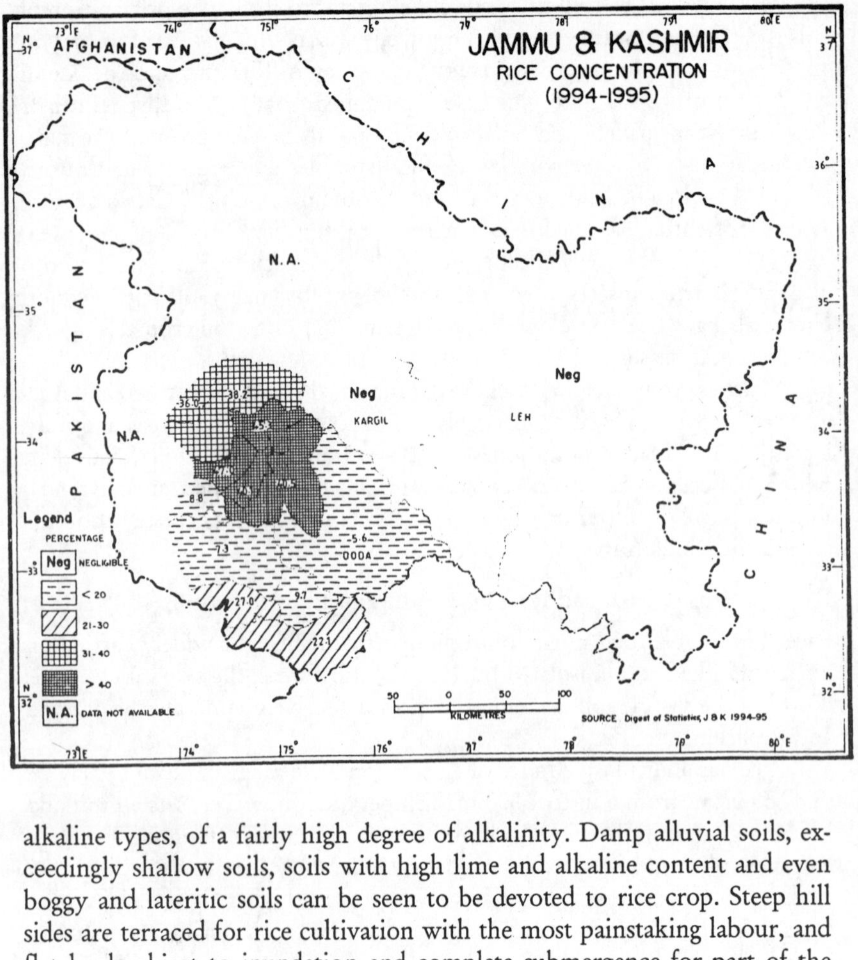

alkaline types, of a fairly high degree of alkalinity. Damp alluvial soils, exceedingly shallow soils, soils with high lime and alkaline content and even boggy and lateritic soils can be seen to be devoted to rice crop. Steep hill sides are terraced for rice cultivation with the most painstaking labour, and flat land subject to inundation and complete submergence for part of the year is also cultivated with rice. The best soils for the cultivation of rice are however, clayey loams of the kind seen on flat land under irrigation and in river deltas. They work into a soft mud when puddled and when dry developed deep cracks.

The cultivation of rice is carried out in many ways but the most important preliminary is the laying out and levelling of the fields for the purpose. Cultivation of rice requires that water should stand evenly over the field practically throughout the growing period and be capable of flowing out in gentle flow, the level of the water in the field being kept up at the same time. Fields, therefore, have to be level and bounded by firm, fairly watertight bunds. The size of the fields has to be suitable, both for the purpose of impounding the water at an even depth and to permit a smooth flow. On sloping ground and undulating country or the sides of the hills, the difficulty and expense of making any deep cutting and somewhat large terraced fields sets on additional limit to the size of the fields. Fields should be laid out as far as possible into suitable rectangles and as large as the surface of the tract will allow, consistent with the requirements of irrigation.

Raising Nurseries

Seedlings of rice are raised in a separate highly manured nursery. The nursery is prepared either 'wet' or 'dry'. For the nursery, the soil of a small selected field is prepared in the same way as the puddle cultivation, i.e., the field is flooded, ploughed in puddle, and this process is repeated until a very fine puddled condition is obtained. Sprouted seeds are sown at the rate of about 10 kg per 1/40 acre or about four quintals per acre in the soft mud with about three cm of water. Irrigation of the seedling nurseries continued without a break until the 30th day when the seedlings are usually pulled out for transplantation.

Seedlings are also raised by dry method. In the dry nursery, plot is given a soaking irrigation to soften it, and when it has dried slightly, it is ploughed or dug thoroughly and prepared clods are broken, weeds gathered and removed the ground is well harrowed and brought to a fine tilth; it is laid into narrow beds about three metres wide and separated by furrows, the beds being slightly elevated. Cattle manure is spread on the beds and well worked in. Subsequently, seed is sown and harrowed or stirred in into the beds and another layer of cattle manure is spread on the surface. The surface is kept moist by irrigation. After the seeds sprout and are well established, watering is given once in two or three days.

Rice is a soil exhaustive crop. The rice growing fields are given adequate manures to replenish soil fertility. The ordinary manures used in rice cultivation are cow-dung, ashes, night-soil, green manures, and chemical fertilizers like ammonia, phosphates, nitrates (NPK). In the Valley of Kashmir, penning of sheep is a common practice. Nursery beds may receive sheep manure which enhance the soil fertility substantially.

Harvest of rice generally begins in the months of September and October. After the crop is harvested, it is allowed to lie in the fields for some time upto three to four days, and then brought over to the threshing ground. The threshing starts after about 10 to 15 days when all the crop has been harvested. The threshing is carried out by beating the earheads on an inclined plank or a bench or stone, the grains separate with great ease though in many cases a small proportion is left over in the straw. Some of the farmers trample out the grain under the feet of cattle. The grain is then winnowed to remove chaff or empty strains of which all bulk grains always contain more or less. The average yield of rice varies from macro to meso and micro-regions. In general, the average yield of rice in Kashmir Division is about 15 quintals per hectare while in the Jammu Division, the average yield per hectare is about 13 quintals. In both the cases, it is above the national average of 12 quintals per hectare. Out of the total production of rice in 1996-97, about 78 per cent was produced in the Kashmir Valley and the remaining 22 per cent in Jammu Division. The productions of rice may be enhanced by diffusing new seeds of the High Yielding Varieties, developed by the Indian Council of Agricultural Research and by providing the recommended quantity of water, fertilizers and insecticides and pesticides.

Wheat (Triticum Sativum): *Kanak*

Wheat is the staple food in the Jammu Plain, adjacent to the province of Punjab. Wheat is a crop which can be grown in the tropical and subtropical, temperate and cold zones even beyond 67°N. Generally sown in autumn, it can stand the cold of the severe winter months and resumes growth with the advent of warm spring and summer. In altitude also, wheat is remarkably adaptable and its cultivation may be done upto 3,000 m above the sea level. At the time of sowing an average temperature around 20°C is conducive for its germination, while at the ripening time, the average temperature should vary between 25°C to 27°C. Its cultivation requires about 30cm of rainfall well distributed over the four months of growth period or there should be assured supply of water by canals and tubewells.

Wheat thrives well in the well-drained alluvial soil, uniform in texture, and free from stones, gravels, and coarse sand. Wheat also requires a well-prepared soil in good tilth and thoroughly free from weeds. In Jammu and Kashmir Division, it is sown in the months of October and November, while in Ladakh (Suru and Nubra valleys) it is a Kharif crop, sown in the month of May and harvested in the middle of August.

The field of wheat has to be ploughed several times, clods broken, and all weeds and stubble removed, and burnt. The field is to be manured with cattle manure. The soil is ploughed and rolled several times to obtain a fine tilth and a firm seed-bed. Seed is sown generally in shallow furrows drawn about 20 cm apart, the seed being dropped through one-furrow drill. About 40 to 50 kg of seed is required for an acre. The field is lightly harrowed to cover the seed, and is left level for forming irrigation beds. Wheat is irrigated after three weeks from the date of sowing and the new seeds of High Yielding Varieties require watering after every 15 days. Wheat responds strikingly to chemical fertilizers (NPK). The crop is harvested after about 110 to 120

FIGURE 8.5

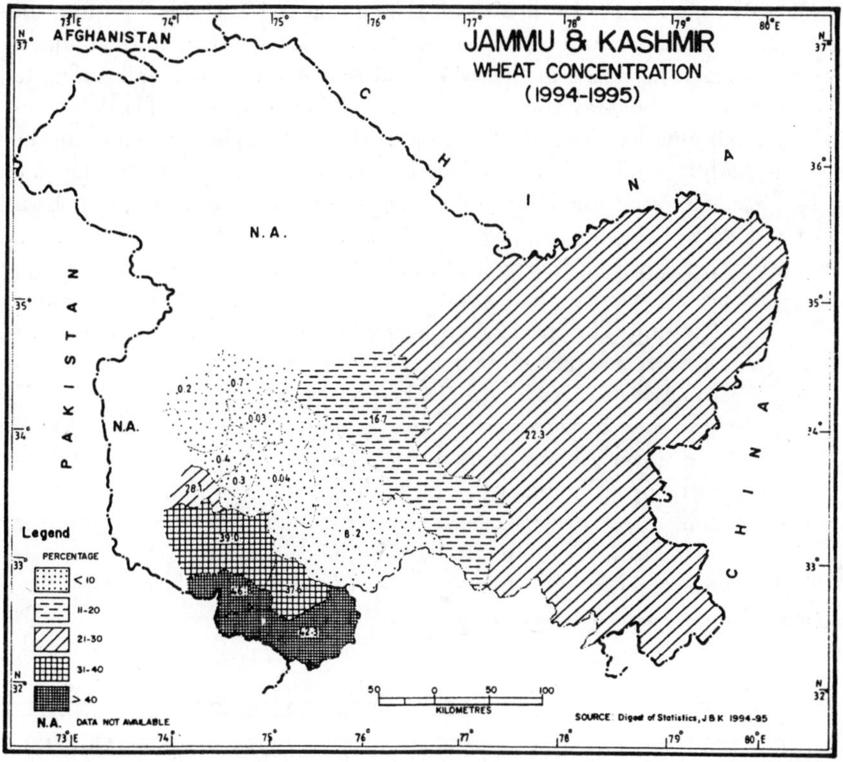

days. The harvesting is done with sickles. The sheaves are brought to the threshing floor and stacked, and the grain is trashed, generally in thrashers after one week depending on the weather condition.

The High Yielding Varieties of wheat are highly susceptible to pests and disease. The crop is often affected adversely by rust of wheat, black, brown and yellow. When the crop in attacked by rust the whole crop is ruined. The disease is seen in the form of yellow/black rusty spots on the leaves, asepti-cally when the crop is well grown, which extend and eventually dry up the crop. Now some new varieties, resistant to rust have been developed by the Pusa Institute, New Delhi.

Wheat is also subject to 'smut' and 'bunt' in the same way as other cereals, the affected grain reduces to a mass of black powder. The treatment for smut and bunt diseases consists in subjecting the seed to spor-killing agencies like hot water, formalin and copper sulphate.

Wheat has been diffused in almost all the district of the state. The spatial distribution of wheat has been shown in Fig. 8.5. Its major concentration is however, in the Jammu (42%), Kathua (42%), Udhampur (37%), and Rajauri (39%). In the rest of the districts of the Jammu and Kashmir Division, the terrain and temperature conditions are not conducive for its cultivation, except in small hectarage in the valley and river terraces. It is interesting to note that wheat has been successfully diffused in the Kargil and Leh districts of Ladakh Division. In Ladakh, wheat cultivation is being carried on in the Suru, Nubra and Rapsu valleys, where it is sown in the month of May and harvested in August and September. On the whole wheat occupies about one fourth of the gross cropped area in the state.

Wheat cultivation was done in about 212,000 in 1996-97, out of which 205,000 hectares were in the Jammu Division and about 4,000 hectares in the Kashmir Division. The Suru Valley (a tributary of the Indus) is coming up as the 'Wheat Bowl' of Ladakh in which an area of over 2,000 hectares was under wheat in 1996-97. In 1964-65 just before the diffusion of High Yielding Varieties, the total production of wheat in the state was 11.2 lakh quintals which rose to over 20 lakh quintals in 1981-82 and 34.75 lakh quintals in 1992-93. In 1996-97 the total production of wheat in the state was about 40 lakh quintals. The increase in the area and production of wheat shows that farmers are increasingly more interested in the cultivation of this crop. The district-wise areal strength of the major food crops (rice, wheat, maize) has been given in Table 8.4.

Table 8.4

Jammu and Kashmir: District-wise Distribution of Cereal Crops, 1994-95

	District	*Rice*	*Wheat*	*Maize*
1.	Anantnag	51.20	0.50	26.65
2.	Pulwama	55.40	0.45	21.82
3.	Srinagar	55.50	0.05	21.75
4.	Budgam	53.60	0.48	23.00
5.	Baramulla	39.00	0.75	28.40
6.	Kupwara	36.75	0.20	51.11
7.	Leh	—	23.00	—
8.	Kargil	—	18.00	—
9.	Jammu	54.35	47.80	20.00
10.	Udhampur	15.50	37.6	69.00
11.	Doda	6.90	8.2	81.65
12.	Kathua	45.75	42.3	32.00
13.	Rajauri	12.75	39.0	82.35
14.	Poonch	4.40	28.0	28.00

Source: Digest of Economics and Statistics, Jammu and Kashmir, 1994-95.

An examination of Table 8.4 shows that rice occupies over 50 per cent of the total cropped area in Anantnag, Pulwama, Srinagar, Budgam, and Jammu districts. It occupies between 30 to 50 per cent in Baramulla, Kupwara and Kathua districts, while in the remaining districts its areal strength varies between 4 per cent (Poonch) to 15 per cent in Udhampur. Leh and Kargil are the only two districts of the state in which rice is not grown at all.

Wheat has a high concentration in Jammu (48%), Udhampur (38%), Rajauri and Poonch districts 42 and 39 per cent respectively. In Kargil and Leh it occupies 18 and 23 per cent of the total cropped area respectively. It is interesting to note that in Leh and Kargil districts the farmers adopted the High Yielding Varieties of wheat which are sown in the months of May and harvested in August and September.

In addition to rice, wheat and maize, a number of coarse grain crops are also grown. Among the coarse grains, *jowar*, *bajra*, small millets, *Kodon* and barley are important. Barley cultivation is confined to Jammu Division and Ladakh. Bajra cultivation is concentrated in Jammu Division, its main growers being Jammu and Kathua districts.

Other food grains include pulses. Pulses are grown in almost all the districts of the state. About 34.25 thousand hectares or 3.58 per cent of the total cropped area is under their cultivation. Fruits and vegetables occupy about 6.5 per cent of the gross cropped area, while oilseeds, condiments, saffron, spices and other food and non-food crops collectively occupy about one per cent of the total cropped area.

Saffron (Crocus Sativus), *Kong*, Zafran, *Kesar*

Saffron is the slender, dried, redish-brown, flattened stigma of cultivated saffron plant (Crocus Sativus). It is an important and expensive condiment used in medicines and foods. It adds a faint, delicate and pleasing flavour to food and special dishes. It has sedative properties and been considered as a panacea for numerous diseases of males and females. It is also used for perfumes. At present, its maximum consumption is for the preparation of *Pan-Masala* (a chew).

Crown in the Karewas of Kashmir and Bhadarwah (Jammu Division), it in an important cash crop, providing employment to about 5 per cent of the total rural workforce in the Valley of Kashmir.

Saffron is being grown in the Pampore karewa of Kashmir since the time immemorial. It is not known when the cultivation of saffron was started in Kashmir, but saffron is being reported to have been an important ingredient of the prescriptions of Veghbhatta and Sustratta who practised medicine about 500 B.C.[1], Its cultivation was in vogue in Pampore (Padampure) even when Kalidas wrote his literary masterpiece *Shakuntala* and

Meghdoot. Kalhan, the well-known historian of Kashmir stated in *Rajterangani* that saffron was under cultivation in Kashmir even before the regime of Lalitadatiya during 725 A.D.[2]. Abul Fazal in his *Ain-i-Akbari* says that the saffron fields in blossom afford a prospect that would enchant those who were most difficult to please.[3] Despite all these stories and beliefs, it is not clear when the saffron cultivation was started in the Valley of Kashmir.

Saffron is a perennial crop, lasting for about 10 to 15 years after sowing. Its cultivation demands a lot of labour and capital in the first year. The agricultural operations involve, decision-making, selection of plot, purchase of expansive seeds sowing, digging, pulverization, hoeing, harvesting, post-harvest care and disposal in the local, national and international markets.

Saffron in the valley is being grown in the karewa as well as in other categories of loamy soils. The lifecycle is however, longer in the karewas. Apart from financial constraint for the purchase of expensive seeds, the farmer avoids the fields in the vicinity of large trees (poplar and willow) as they reduce the moisture content in the soil. Moreover, the plot should not be close to any waterbody as percolation and creeping of water may damage the seeds of the crop. For better returns, the underground watertable in saffron growing areas should range between 2.5 to 7.5 m.

Once a farmer takes a decision for sowing saffron, the preparation of land begins at least one year before the sowing of corms. It is believed by the farmers of Pampore karewa that sowing of oilseeds in the preceding year is beneficial for saffron sowing in the subsequent year. After fallowing the land for about six to eight months, the actual preparation of land for sowing begins with the advent of *Sonth* (spring season). The field is ploughed in the month of March or April, depending on the weather conditions. Ploughing of field with more moisture is not advisable. Subsequently, about 40 to 50 tonnes of farmyard manure is applied per hectare. The field is ploughed again in the month of May to obtain a fine tilth. The farmers, on the basis of their empirical experience attach special importance to the ploughing of field in the last week of June as it helps in the eradication of weeds. The ploughing of field continues after every 15 days till the end of August. The agricultural operations involved in saffron cultivation have been shown in Fig. 8.6.

The usual time of planting of corms commences from the last week of July to the last week of August. The second fortnight of July is, however, considered as the best period for the sowing of saffron corm. A delay in the planting of corm has an adverse affect on the lifecycle of corm. The corms selected for planting has a marked influence on the size and production of flowers and their stigmas. The seed corms should be healthy and free from fungal diseases. The elderly farmers of the Pampore karewa opine that the

FIGURE 8.6

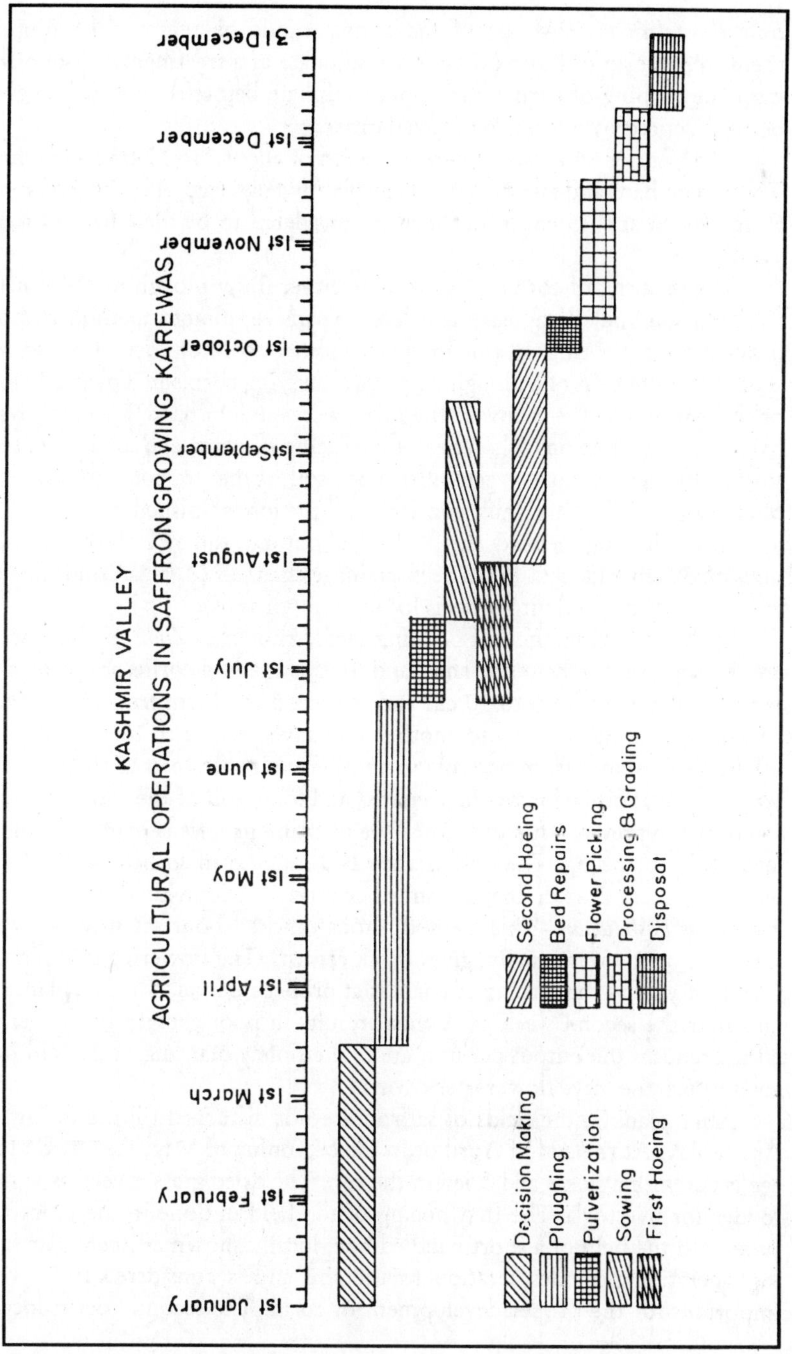
KASHMIR VALLEY
AGRICULTURAL OPERATIONS IN SAFFRON GROWING KAREWAS
Ist January
Ist February
Ist March
Ist April
Ist May
Ist June
Ist July
Ist August
IstSeptember
Ist October
Ist November
Ist December
31 December
Decision Making
Ploughing
Pulverization
Sowing
First Hoeing
Second Hoeing
Bed Repairs
Flower Picking
Processing & Grading
Disposal

outermost fibrous covering of the corms should be removed by dipping them in solution of 5 per cent copper sulphate as a treatment before planting. The dipping of corms into copper sulphate before sowing is however, being practiced by a few progressive farmers.

The seeds of saffron are sown at a depth of about 7 to 10 cm (3-4 inches). There is no hard and fast rule about the planting distance. A spacing of about 5 cm within and 15 cm in between is considered to be ideal for optimum yield[4].

The planting of corms is being done either (i) by plough method or (ii) by *Zoon* method. Being easy and least expansive, plough method is more prevalent for planting. In this method, about 7 to 10 cm deep furrows are made with the help of a plough. A person carrying corms in a piece of cloth or in *Pharan* puts the corms in the furrow on random basis. He is followed by one or two persons who check the spacing of corm and advise the seed sower, after every furrow gets filled with soil of the previous furrow. The ploughing and planting, thus, continues till the job is finished.

After planting the seed, the field is laid out into square beds of about 2.5 m, with 30 cm wide and 15 cm deep drainage channels (*Athe*) around. The division of saffron field into units is locally known as *Skust*.

In the second method of planting seeds, known as *Zoon* method, seeds are shown with the help hoe and hand. In this method corms are planted in furrows at a depth of 8 to 10 cm and spacing 5 to 10 cm from each other. The distance from one row to another varies between 12 to 18cm from each other. The distance from one row to another varies between 12 to 18cm. After planting another furrow in prepared and loose soil of the furrow is used to cover the previous furrow. Once the planting process is over, the field is levelled by planking. The field is then laid into 2.5 m sq beds with 30 cm wide and 15 cm deep drainage channels around. The individual saffron bed is known as *Poshawaare.* This method, though more labour intensive as every seed is to be sown manually, gives quick returns. The flowering starts in the very first year as the placing of corm is at proper interval. The late planting, i.e., after the second week of August, results in poor growth by flowering which reduces the output per unit area. The timely planting of seeds in July gives return, i.e., five flowers per corms[5].

After planing, the fields of saffron are not disturbed till the following April/May. At the end of April or in the beginning of May, the saffron foliage is cut with a sickle and dried in the sun. The dried grass is used as winter fodder for livestock. The first hoeing of the field in done in the month of June with the help of a short handled hoe, locally shown as *Zoon.* The hoeing operation provides aeration to the soil and is considered to be very important for the proper development of corm. The second hoeing locally known as *Sorue*, is generally done in the first week of August which is considered to be highly beneficial for higher yields. The third and the last hoeing is done in the month of September just about 30 days before flowering.

Once the third hoeing in over, the operation of bed repairing starts. The drainage channels are cleared in order to save beds from waterlogging. Subsequently, the soil is pulverized with a wooden piece, called *Yethapur* till the clods are pulverised to power. This improves aeration through which the delicate stems of flowers emerge on the surface.

The lifespan of a saffron crop largely depends on the texture and moisture content of the soil. The general prevailing temperature and moisture at the time of sowing influence the growth of the crop, while after a week or 10 days, a light shower enhances the proper germination. As stated above the lifespan of saffron in karawa lands is generally 10 to 15 years while in the non-karewa soils it is only 5 to 7 years.

The flowers start appearing in the third week of October. From 20th October to 15th November is the flowering period. This is the period when the family labour remains overoccupied. Many of the farmers engage hired labour at high wages, ranging from Rs. 75 to 100 per day. The plucking of flowers starts in the morning around 10 a.m. after the disappearance of dew. Plucking of flowers is an art which demands great skill and dexterity. Plucking of flowers is to be done only with fingers without trampling the soil below. Saffron flowers are collected in specially designed baskets of bamboo and willow. The job is quite demanding as about 150,000 flowers produce about one kg of saffron. The plucking of flowers has to be done by standing in the drainage channel and bending oneself to pick the flowers by stretching hand. In this process, extra care is to be taken so that the budding flowers do not get bruised and damaged.

The time gap between the successive plucking depends upon the tradition and circumstances of the saffron growing areas. In the Pampore karewa, the picking of flowers is done on alternate days, while in Pyre and Badgam karewas flowers are plucked after every two days. After about a fortnight from the date of first plucking of flowers, the number of flowers per unit area diminishes. Around 15th of November the brief season of harvest comes to an end.

Diffusion and Distribution of Saffron

Saffron cultivation in Kashmir has a long history. Reliable data on its area, production and yield for the historical period has however, not been available. Saffron cultivation after 1960 has been diffused in several non-traditional areas outside of Pampore (Fig. 8.7). At present, it covers an

FIGURE 8.7

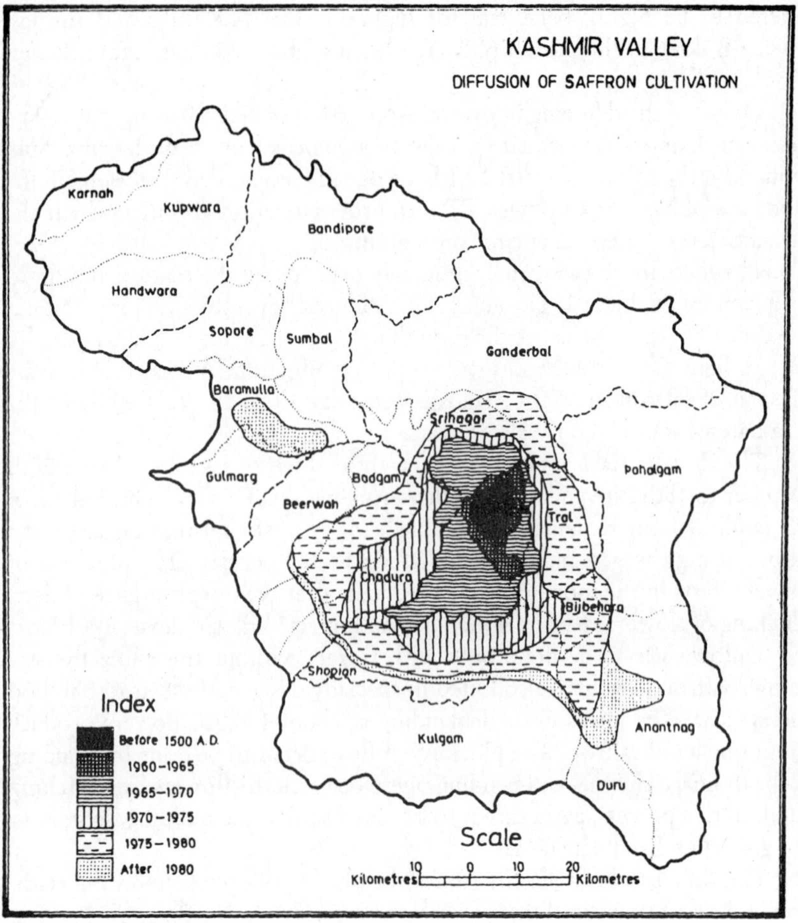

area of about 4,466 hectares which is about 1.17 per cent of the gross cropped area and about 4.50 per cent of the area under commercial crops in the Valley of Kashmir. The temporal growth in area, production and yield has been given in Table 8.5.

It may be seen from Table 8.5 that within a period of 15 years (1980-95) the area under saffron has increased by about 72 per cent over the base year of 1980. It is a significant increase in the areal strength of saffron and validates the fact that saffron is emerging as the leading cash crop in the karewas of Kashmir. The total production of saffron in 1980 was about 98 quintals which went upto about 136 quintals in 1995, thereby recording an increase

Table 8.5

Valley of Kashmir: Growth Pattern in Area, Production and Yield of Saffron Cultivation, 1980-1995

Year	Area in hectares	Production in quintals	Yield Kg/Hect.	% Area	Growth in Production	Growth in Yield
1980	2615.00	98.30	3.75			
1985	3566.00	86.50	2.42	36.40	-12.11	-35.5
1991	4001.00	91.50	1.29	53.00	-6.92	-39.17
1995	4466.00	136.00	3.10	68.55	38.33	-17.55

Source: Digest of Statistics, Directorate of Economics and Statistics, Srinagar.

of 37.70 quintals in a short period of 15 years (1980-95). The highest yield per hectare, i.e., 3.75 kg was recorded in 1980 which fell to only 2.29 kg in 1991 (Fig.8.8). At present the per hectare yield of saffron is about 3 kg (Table 8.5).

The district-wise area of saffron has been given in Table 8.6. So far as the areal concentration of the crop is concerned, Pulwama District has the highest hectarage under saffron cultivation, being 3,415 hectares or about 89 per cent of the total saffron area. The next important district in terms of area of saffron is Budgam. In 1984, about 251 hectares of its area or 6.60 per cent of the total saffron area was in this district. In 1995, the area under saffron in Budgam District was 439 hectare or 9.96 per cent of the total area under saffron in the Valley. The other districts in which saffron cultivation is carried on are Srinagar and Anantnag. In these districts about 210 and 29 hectare were under saffron cultivation in 1995 respectively (Table 8.6).

Table 8.6

Kashmir Valley-Saffron Distribution, District-wise, 1995

District	Year	Area	% of the Total Saffron Area
Pulwama	1985	3415.00	88.91
	1995	3728.00	84.11
Budgam	1984	251.00	6.60
	1995	439.00	9.96
Srinagar	1984	128.00	3.37
	1995	210.00	4.76
Anantnag	1985	4.00	0.10
	1995	29.00	0.65

Source: Computed from the data obtained from the Directorate of Economics and Statistics, Jammu and Kashmir, Srinagar.

FIGURE 8.8

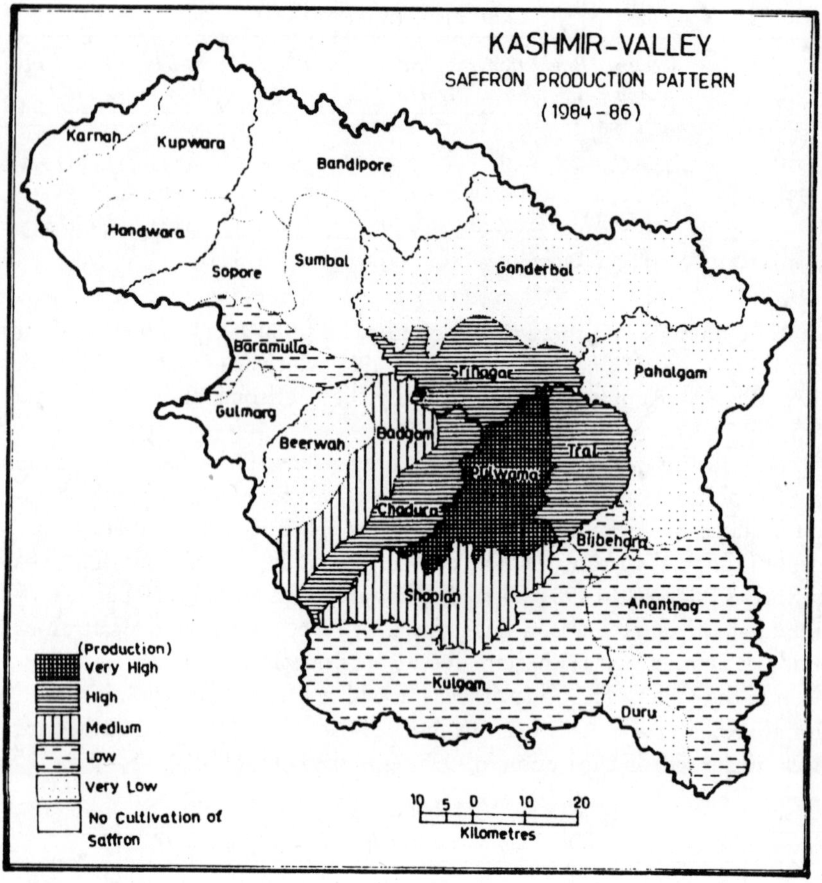

Apart from Pampore, saffron has been diffused in the karewas of Pyer, Khanpura, Nagam, Chadura, Tral, Bejbehara. There are several potential karewa lands in which saffron may be grown successfully. About 400 hectares of land can be brought under saffron in the Zainapora and Bijbehara karewas of Anantnag. Approximately 1000 hectares of land may be brought under saffron in the Cherwini, Hayatpora, Aripal, Nagam, Bandipora and Chadura karewas. Similarly, in the Pulwama District are the Koyul, Ladhu, Kakapora, and Shupiyan karewas which are largely conducive for saffron cultivation. The rainfed karewas of Pattan, Wanigam, Singhpora, Taperwaripora, and Safapora in the District of Baramula are also partly suitable for saffron cultivation.

FIGURE 8.9

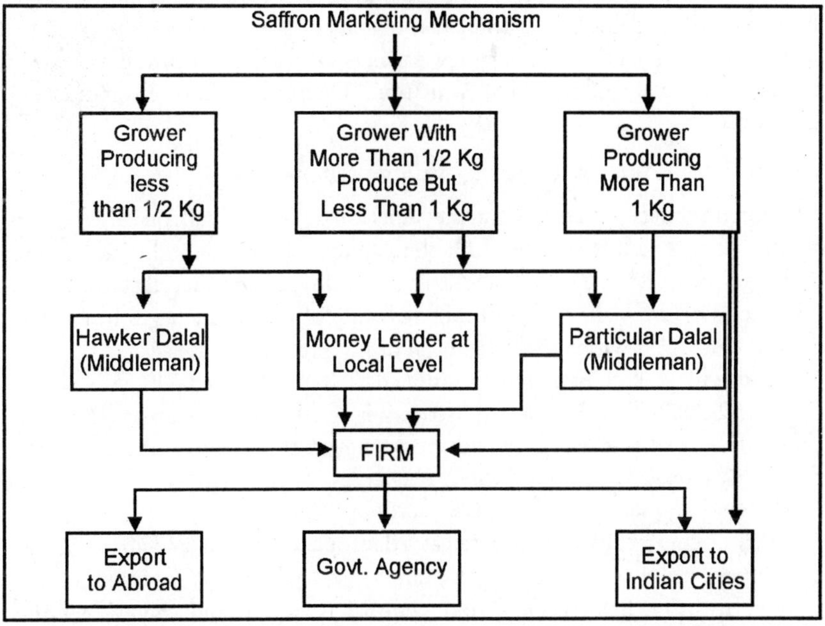

The production pattern of saffron has been plotted in Fig. 8.7. It may be observed from Fig. 8.8 that the District of Pulwama has the highest production and yield in the valley, followed by Chadura, Srinagar and Tral. The Tehsils of Ganderabal, Beerwah, Duru and Pahalgam have very low productivity and production (Fig. 8.8).

Marketing of saffron is largely in the hands of middlemen and their firms. Generally the price of the commodity is determined by the middlemen and the marketing firms. The main marketing channels of saffron in the Valley of Kashmir have been diagramatically presented in Fig. 8.9. An examination of this figure clearly shows the dominance of the private sector in the disposal and marketing of the crop. It may be observed from 8.9 that there is a long chain of middlemen, e.g., *Dalals,* sub-firms, wholesellers, who are operating in the marketing of saffron. These intermediaries buy saffron from the growers who are generally poor and not in a position to dictate terms in the fixation of price of their produce. The middlemen and big firms know the real position of the commodity in the national and international markets. The sum of margin of commission of different middleman results into a low profit to the growers. The small farmers, generally sell their produce to the *Dalals* (middlemen) and sub-firms. The small farmers feel to dispose off their production to *Dallas* and middlemen. Often the small and poor farmers borrow money from the middleman and sell their produce to him to clear the debt. Under these circumstances the farmers sell his produce at a cheaper rate. There are very few farmers who sell their produce directly to the wholesale dealers in Amiritsar, Delhi, Bombay, Calcutta and Lucknow.

Landuse in a Saffron Growing Village Chandhara

Chandhara village is famous for the superior quality of saffron all over the world. Being the birth place of Habba Khatoon (*Zooni*), the literary figure and the wife of King Yusuf Shah Chak, Chandhara has a special status in the chronicles of the Valley of Kashmir. It is one of the rare villages in the history of the world which saw the union of nobility and ordinary.

The village of Chandhara is situated at 33° 59′N and 74° 56′E in the Pampur Tehsil of the Pulwama District (Fig. 8.10). It is only about 17 km to the south-east of Srinagar City and about one km to the north of the Jammu-Srinagar National Highway No. 1. It is accessible by a metalled road. The total area of the village is about 1205 acres or 9640 *Kanals*.

Physiographically, it is situated at a flat topped karewa, and about 70 per cent of its area in covered by *Gurti* (karewa) soil which is well suited for saffron cultivation. Sekil (sandy) and Nambal (swampy and clayey) are the other types of soils, occupying about 9 and 8 per cent of the total area of the village respectively. The general slope of the village is from east to west. The general landuse of the village has been plotted in Fig. 8.11.

Out of the total reporting acres of 1,205 areas about 75 per cent is under cultivation, 9.12 per cent not available for cultivation, and 11 per cent, other cultivated land excluding fallow, while 5 per cent is fallow land (Table 8.7).

Table 8.7

Village Chandhara: General Landuse 1994-95

Land Classification	*Area in Acres*	*Percentage to the Total Area*
I. Land not available for cultivation	109.87	9.12
II. Other uncultivated land excluding fallow	133.25	11.06
III. Fallow land	61.25	5.08
IV. Net area under cultivation	900.00	74.74

Source: After S.S. Shah, 'Growth and Expansion of Saffron Cultivation in Kashmir and its Impact on Economy and Society. Unpublished Ph. D. Thesis, Jamia, N. Delhi, 1998.

An examination of Table 8.8 shows that about 660 acres or about 73 per cent of the net cropped area was under the perennial crop of saffron in 1994-

FIGURE 8.10

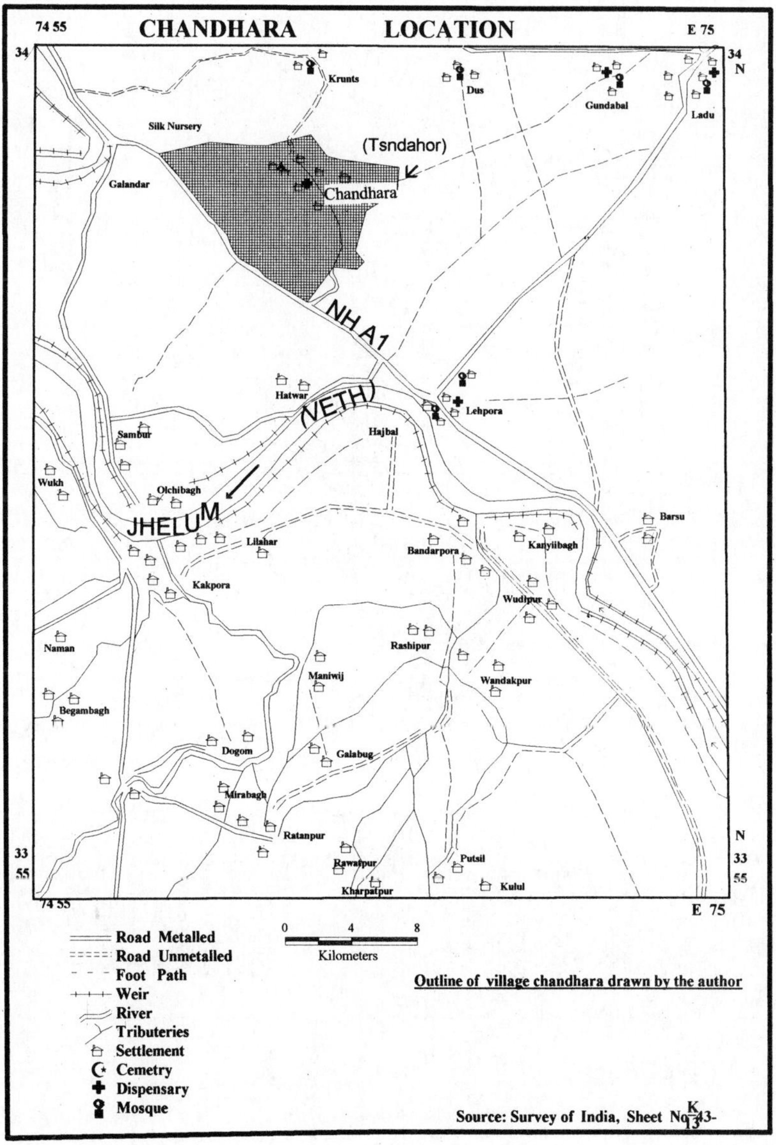
CHANDHARA LOCATION
74 55
E 75
34
N
33
55
Krunts
Dus
Gundabal
Ladu
Silk Nursery
(Tsndahor)
Galandar
Chandhara
NH A1
Hatwar
(VETH)
Lehpora
Hajbal
Sambur
Wukh
Olchibagh
JHELUM
Lilahar
Bandarpora
Barsu
Kanyiibagh
Kakpora
Wudipur
Naman
Rashipur
Maniwij
Wandakpur
Begambagh
Dogom
Galabug
Mirabagh
Ratanpur
Rawatpur
Putsil
Kharpatpur
Kulul
Road Metalled
Road Unmetalled
Foot Path
Weir
River
Tributeries
Settlement
Cemetry
Dispensary
Mosque
0 4 8
Kilometers
Outline of village chandhara drawn by the author
Source: Survey of India, Sheet No 43-K/13

FIGURE 8.11

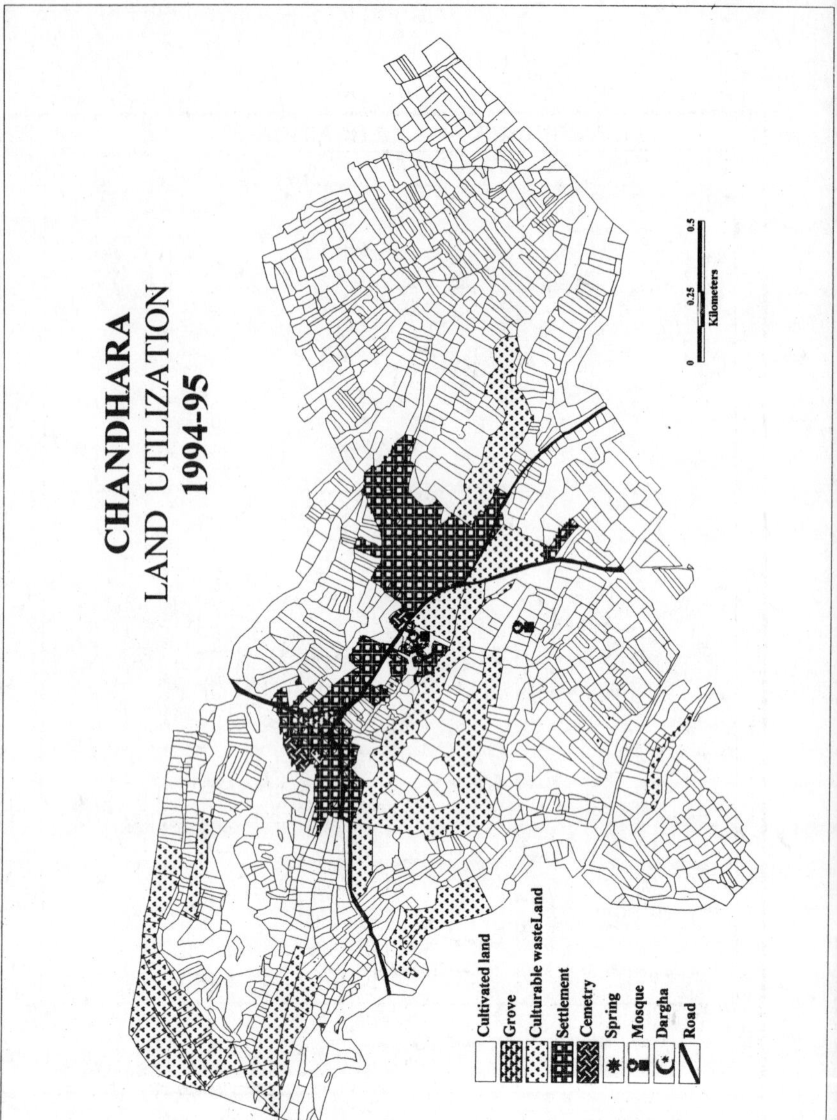
CHANDHARA
LAND UTILIZATION
1994-95
N
Cultivated land
Grove
Culturable wasteLand
Settlement
Cemetry
Spring
Mosque
Dargha
Road
0
0.25
0.5
Kilometers

95. Thus, it was the leading first ranking crop of the village. Almond and orchards occupied 59 and 8 acres, being 6.55 and 0.90 per cent of the net cropped area, while 68 acres or 7.59 per cent was under almond mixed with saffron crop. A new trend of planting saffron in the almond orchards has developed in the village. For the last about 15 years almond orchards have been highly susceptible to pests and disease and not fetching good return to the growers. The farmers opine that mixed almond and saffron is more remunerative than their cultivation as isolated crop. There is an urgent need to develop a marketing system in which the growers may get a reasonable price of their produce. Sooner such a system develops the better.

Rice and maize are the cereal crops grown in the village but they occupy only about 4.43 and 3.19 per cent of the net cropped area respectively, while about 3 per cent area was under vegetables. Over 8.6 acres were devoted to fodder and 6.25 acres were fallow in 1994-95.

Table 8.8

Village Chandhara-Landuse in Kharif Season, 1994
Cultivated Land 961.87 acres
Net Cropped Land 900.62

Crop	*Area in acres*	*Percentage to the gross cropped area*	*Percentage to the net cropped area*
Saffron(perennial crop)	659.38	68.55	73.21
Almond	59.00	6.13	6.55
Orchards	8.12	0.84	0.90
Almond and saffron	68.37	7.11	7.59
Rice	39.87	4.15	4.43
Vegetables	28.50	2.96	3.16
Fodder	8.63	0.89	0.96
Maize	28.75	2.99	3.19
Fallow	61.25	6.37	—
Total	961.87	100.00	100.00

Source: Shah, S.S. "Growth and Expansion of Saffron Cultivation in Kashmir Division..." Unpublished Ph. D. Thesis, 1998.
The land utilization in the Kharif season (1994-95) has been plotted in Fig. 8.12 while Table 8.8 gives the area under different crops.

The rabi landuse patterns have been plotted in Fig. 8.13 and the area under different crops has been given in Table 8.9.

Table 8.9

Village Chandhara: Landuse in Rabi Season 1994-95

	Crop	Area in acres	Percentage to the gross cultivated land	Percentage to the net cropped land
1.	Saffron (perennial crop)	659.38	68.55	77.43
2.	Almond	59.00	6.13	6.92
3.	Orchards (apple)	8.12	0.84	0.95
4.	Almond and Saffron	68.37	7.11	8.02
5.	Fodder (oats)	41.25	4.29	4.84
6.	Wheat	5.25	0.55	0.62
7.	Oilseeds	6.75	0.70	0.78
8.	Vegetables	3.75	0.39	0.44
9.	Fallow	110.00	11.45	—
	Total	961.87	100.00	100.00

Source: After S.A. Shah, 1998.

Saffron, a perennial crop continued to occupy about 77.4 per cent of the cropped area of Chandhara village. Almond and apple orchards occupied about 7 and one per cent of the cropped area respectively in 1994-95. Wheat, oilseeds and vegetables were the other crops grown in the rabi season. Their percentage share in the net cropped area was however, insignificant. Fodder, covering an area of about 4.84 per cent of the sown area is another significant rabi crop in the village. The cultivation of rice is a new phenomenon in the crop land use of the village. For the first time the cultivation of High Yielding Varieties of wheat was started in 1980-81. Wheat still occupies only about 5.25 acres of the net cropped area (Fig. 8.13, Table 8.9).

The area under saffron and its total production have gone up. The yield per unit area is, however, showing a declining trend, especially in the traditionally saffron growing areas, like Pampore. A package programme has to be adopted to enhance saffron production and to make the karewa lands more sustainable. Some of the steps which may go a long way is making saffron cultivation more remunerative are as under:

1. The main investment in the cultivation of saffron is the purchase of seed. The saffron corms are quite expensive. The government should try to provide good quality seed to the growers at a reasonable rate.
2. In the Pampore karewa, saffron is being grown by indigenous technique. The obsolete technology is one of the main causes of soil depletion. Diffusion of innovations and adoption of scientific rotation may help is making saffron cultivation economically more viable and ecologically more sustainable.

FIGURE 8.12

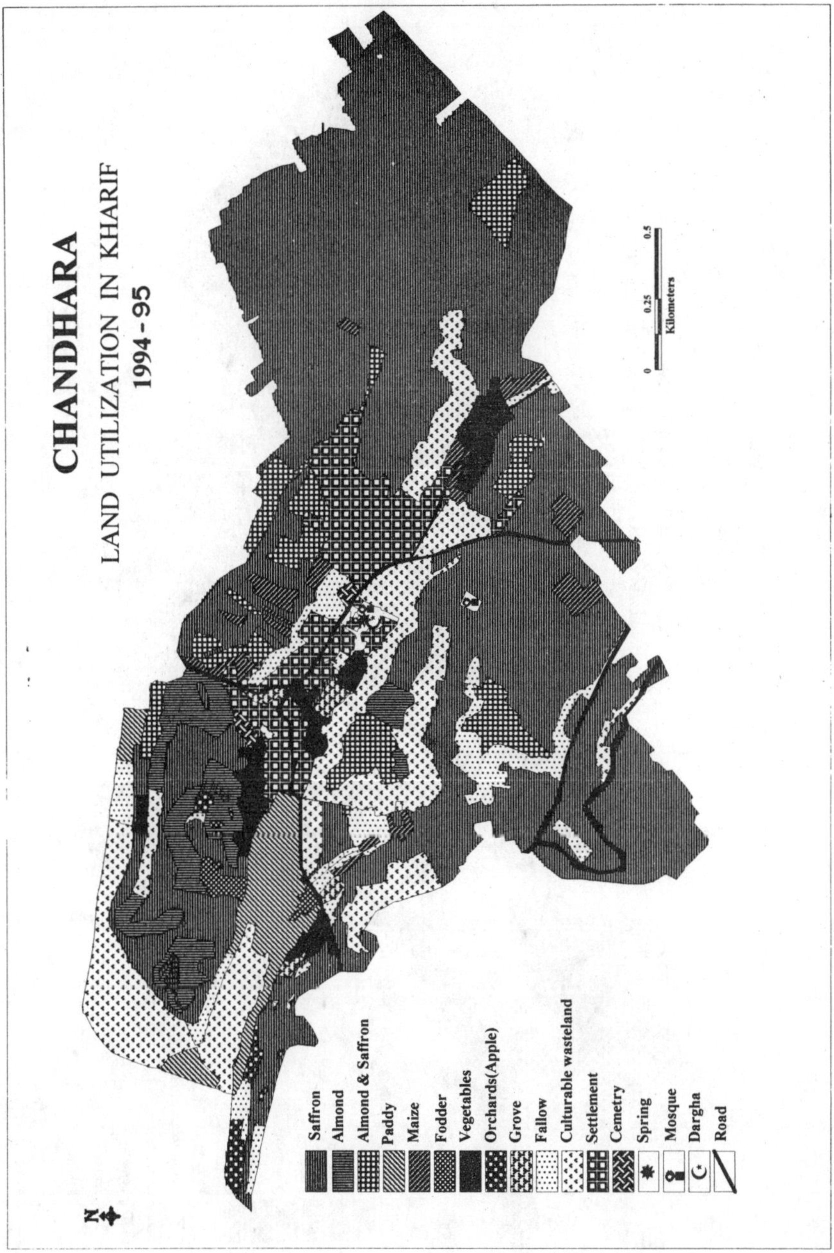
CHANDHARA
LAND UTILIZATION IN KHARIF
1994-95
0
0.25
0.5
Kilometers
N
Saffron
Almond
Almond & Saffron
Paddy
Maize
Fodder
Vegetables
Orchards(Apple)
Grove
Fallow
Culturable wasteland
Settlement
Cemetry
Spring
Mosque
Dargha
Road

FIGURE 8.13

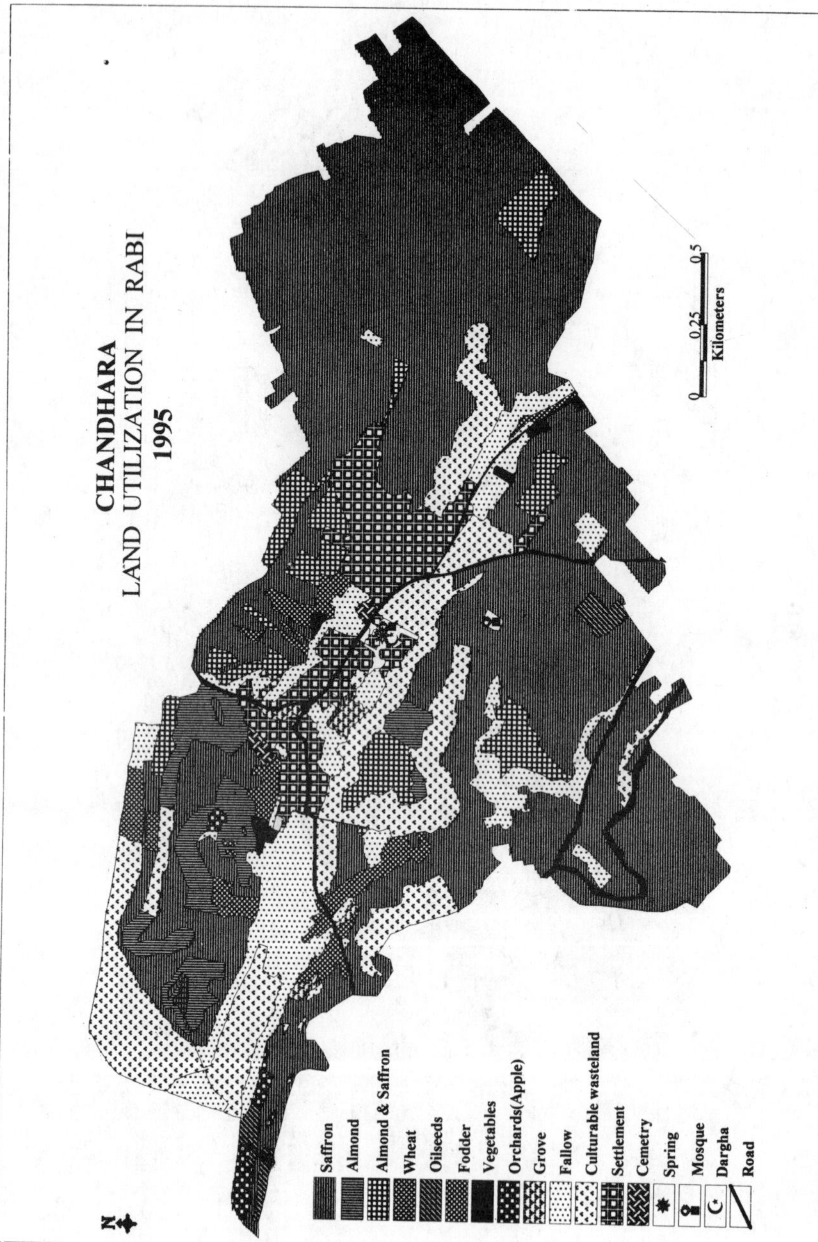
CHANDHARA
LAND UTILIZATION IN RABI
1995
N
Saffron
Almond
Almond & Saffron
Wheat
Oilseeds
Fodder
Vegetables
Orchards(Apple)
Grove
Fallow
Culturable wasteland
Settlement
Cemetry
Spring
Mosque
Dargha
Road
0
0.25
0.5
Kilometers

3. Plantation of social forestry in the vicinity of saffron fields and along the roads should be discouraged as the planted trees of poplar and willow reduce the moisture of saffron fields. Consequently, the growth of crops is adversely affected. This aspect deserves immediate attention of the cultivators and planners.
4. In recent years intercropping of saffron in almond orchards is increasing. The saffron crop, however, does not thrive well under the shadow of large trees. Moreover, the roots of almond trees create problems for the tender corms of saffron. This practice of intercropping either should be discouraged or some new varieties of almond with less foliage need to be developed.
5. A chemical analysis of the Pampore and other saffron growing karewas is necessary to determine the elements which are deficient in the soil. Subsequently, the deficient elements may be added to the respective soil in the form of chemical fertilizers to enhance the fertility of the soil.
6. The saffron growers feel shortage of labour at the peak labour demand. The digging and hoeing operations are quite arduous. A suitable technology needs to be developed to replace manual digging and hoeing by machine. Similarly, the processing of stigmas is a time consuming manual exercise. The female members and children of the family process the stigmas from the flowers with great skill and dexterity. A suitable technology is to be developed to reduce the burden of work on the family.
7. The old varieties of saffron corms are highly susceptible to disease. The Indian agricultural scientists should develop some new seeds which may be more sturdy and resistant to diseases.

If all the given steps are taken together, the yield and production of this valuable cash crop may go up substantially and the karewa soils would become more sustainable.

Ranking of Crops

In any scheme of agricultural regionalisation, the ranking of crops has great significance. The ranking of crops not only gives an idea about dominant crops of a region, it also helps in knowing the crops which are competing for area with each other. The first ranking crops of the state of Jammu and Kashmir have been plotted in Fig. 8.14.

An examination of Fig. 8.13 shows that rice is the dominant crop in the Kashmir Division excepting the District of Kupwara in which it is replaced by maize. The districts of Anantnag, Baramulla, Badgam, Pulwama and Srinagar are the leading producers of rice in the state.

Maize is the first ranking crop in the Doda, Rajauri, Poonch and Udhampur districts (Jammu Division), while wheat gets first rank in the

FIGURE 8.14

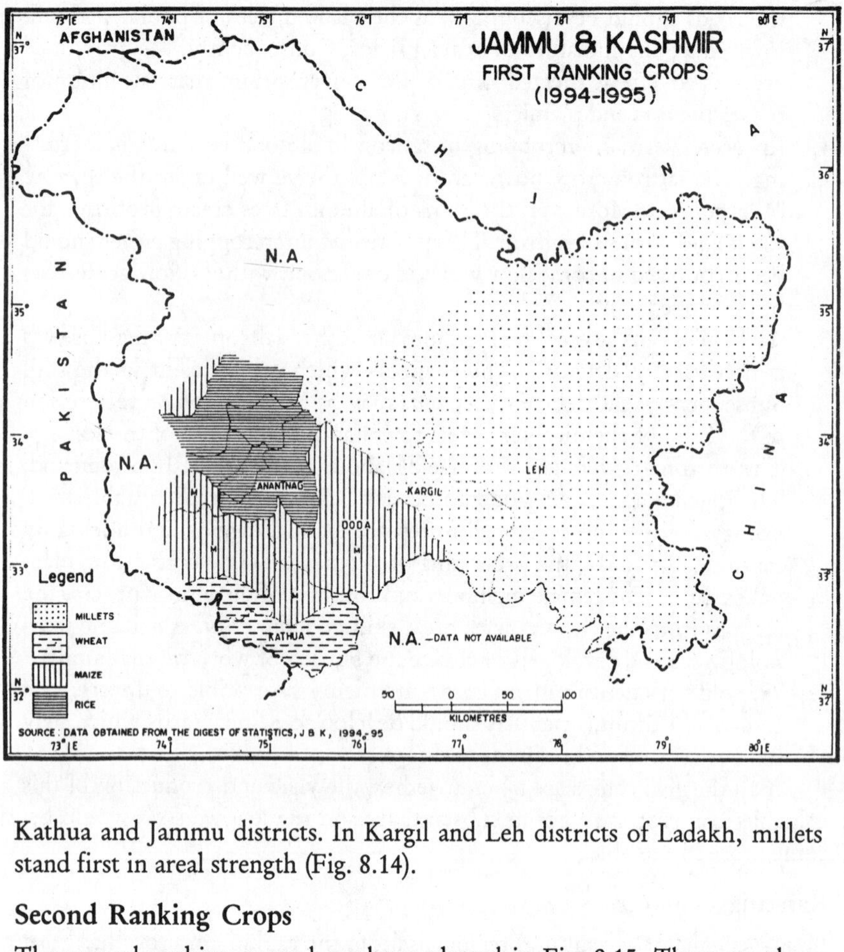

Kathua and Jammu districts. In Kargil and Leh districts of Ladakh, millets stand first in areal strength (Fig. 8.14).

Second Ranking Crops

The second ranking crops have been plotted in Fig. 8.15. There are three crops namely, maize, rice and wheat which get second rank in the various districts of the state.

It may be observed from Fig. 8.15 that in all the districts of the Kashmir Division excepting Kupwara, maize is the second ranking crop. In Kupwara rice gets second rank in area. In the district of Kathua and Jammu, rice is the second ranking crop, while in the remaining districts of the state wheat stands second (Fig. 8.15).

Crop Combinations

Crops are generally grown in combination, and it rarely happens that a

FIGURE 8.15

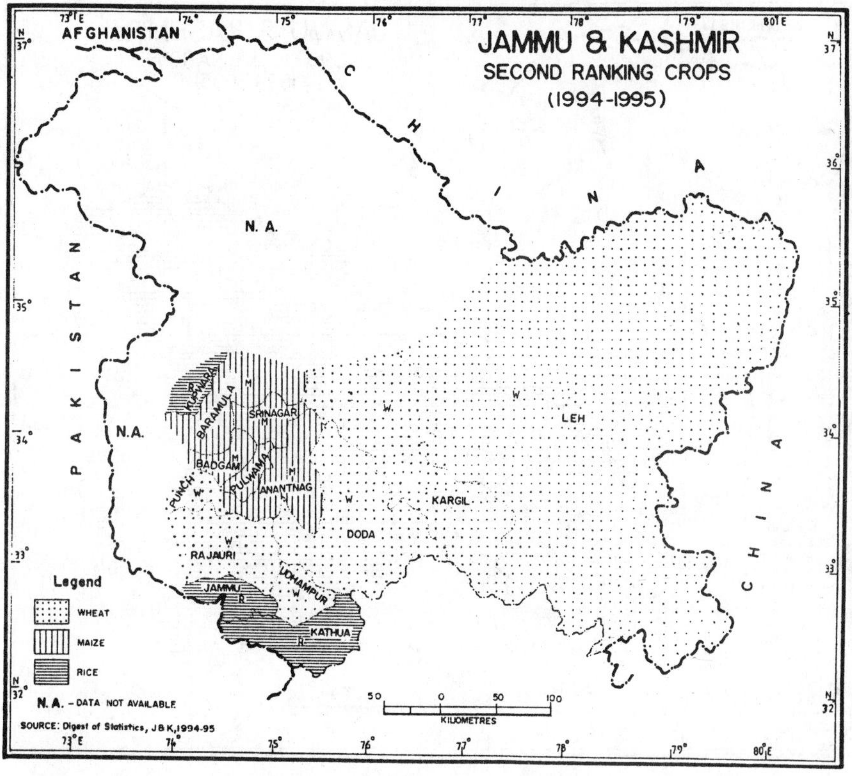

particular crop occupies a position of total isolation from other crops in a given region at a given point of time. The distribution maps of individual crops are interesting and useful for planner and policy-makers, but it is even more important to view the integrated assemblage of the various crops grown in a region. For example, the demarcation of Jammu Division into wheat region does not explain the agriculturally significant fact that the region under question has also rice and maize crops. For a comprehensive and clear understanding of the cropping mosaic of an agricultural region it is imperative to delineate crop combination regions. .

For the demarcation of crop combinations regions of the Jammu and Kishmir state, the standard deviation technique, advocated by Weaver was adopted and the resultant crop combinations have been plotted in Fig. 8.16.

It may be seen from Fig. 8.16 that all five crops (maize, rice, wheat, millets, oilseeds) enter into combination. It may also be observed from the figure that monoculture is not practised in the state. The given crops make

FIGURE 8.16

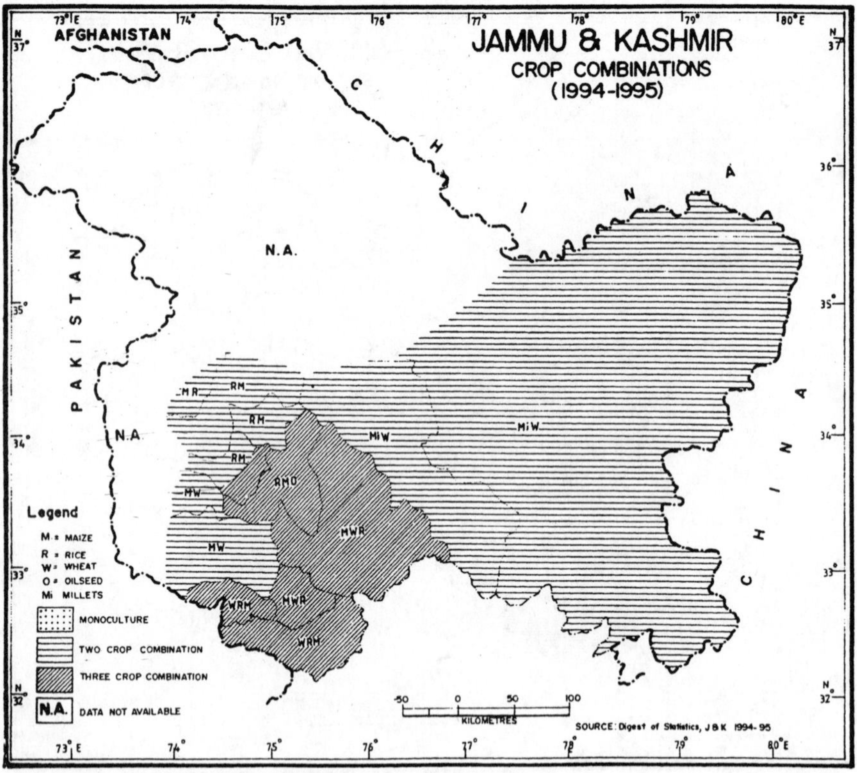

only two and three crop combinations (Fig. 8.16).

The two-crops combination is found in the Baramulla, Badgam, Kupwara and Srinagar districts of Kashmir in which rice and maize are the constituent crops. In Kargil and Leh districts of Ladakh, millets and wheat are making two-crops associations.

The three-crops combination is found in six districts, namely, Anantnag, Baramula, Ganderbal, Badgam, Srinagar and Pulwama of the Kashmir Division in which rice, maize and oilseeds enter into combination, and Kathua, Jammu and Udhampur of the Jammu Division. In the Districts of Jammu Division, wheat, maize and rice are the constituent crops of three crop combination (Fig. 8.16).

As stated in the preceding paragraphs, the cropping patterns and crop association in the various districts of the state are closely influenced by the terrain, soil and extreme climatic conditions. In fact, in the Valley of Kashmir and in the hilly districts of the state, severe winters do not permit the cultivation of rabi crops. Consequently, the farmers are not diversifying

FIGURE 8.17

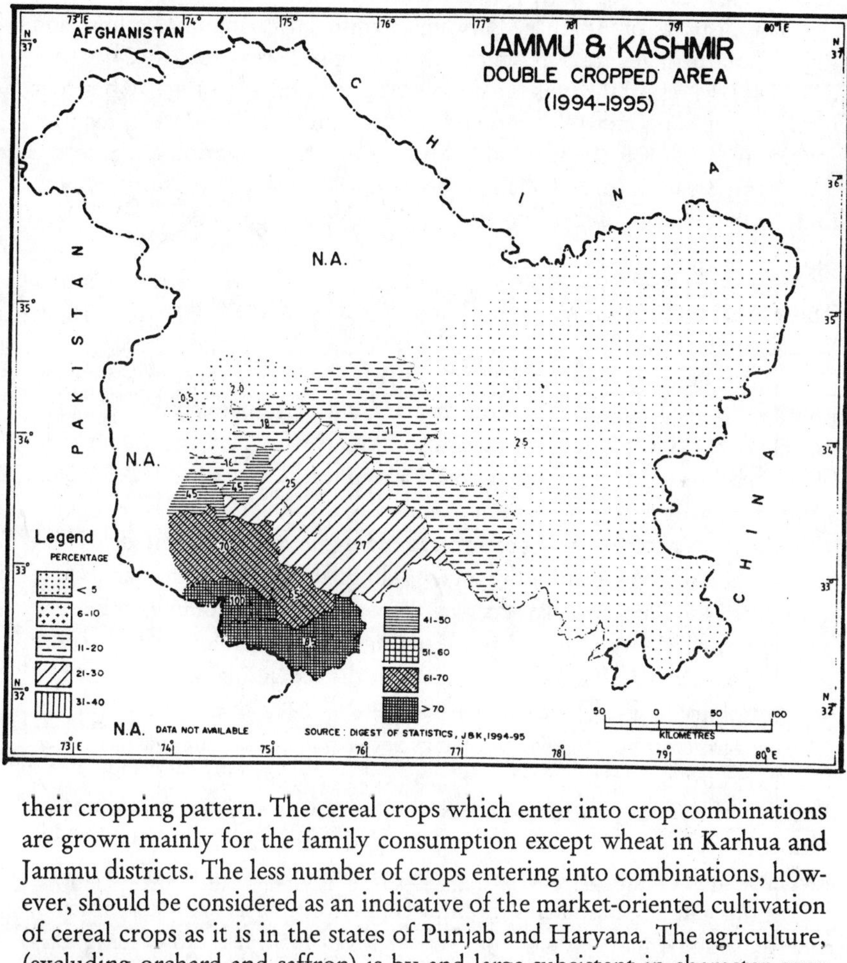

their cropping pattern. The cereal crops which enter into crop combinations are grown mainly for the family consumption except wheat in Karhua and Jammu districts. The less number of crops entering into combinations, however, should be considered as an indicative of the market-oriented cultivation of cereal crops as it is in the states of Punjab and Haryana. The agriculture, (excluding orchard and saffron) is by and large subsistent in character over the greater part of the state.

Double Cropped Area

Out of the total area sown (1074 thousand hectares) about 730 thousand hectares or 68 per cent was double cropped area in the state in 1994-95. The undulating and mountainous topography, harsh winters and inadequacy of irrigation in Ladakh and Kandi tracts are the main obstacles in the intensification of agriculture. Consequently, the double and multiple cropped area is largely confined to the plain areas of Jammu Division and the valley floor of the Kashmir Division (Fig.8.17). In Ladakh, cultivation of crops is possible in only those tracts where irrigation facilities are available.

The districts of Anatnag, Pulwama, Jammu, Udhampur, Kathua, and Rajauri are the main areas in which double cropped area exceeds 70 per cent (Fig. 8.17). In rest of districts the double cropped area is less then 10 per cent which shows a poor state of agricultural development. The development of irrigation in the Kandi areas and cultivation of short duration crops in the Valley of Kashmir may enhance the double cropped area substantially, thereby making agriculture a more remunerative occupation.

Size of Holdings

The size of holdings of the Jammu and Kashmir state at the district level has been given in Table 8.18.

It may be observed from Table 8.18 that about 74 per cent of the total holdings in the state are below one hectare and 25.14 per cent are between one and five hectares. There are only 225 farmers who have over 20 hectares of land each. The large holding farmers are confined to the hilly and mountainous tracts of Anantnag, Jammu, and Leh.

The district-wise analysis of size of holdings show that in the district of Srinagar about 95 per cent of the total holdings are below one hectare, while in Budgam and Kupwara districts over 85 per cent of the holdings are below one hectare. The size of holdings in Baramulla District is also very small as about 81 per cent are below one hectare. In the remaining districts 54 to 78 per cent of holdings are below one hectare.

The farmers who are having holdings between 1 to 5 hectares vary between 13 per cent in Pulwama District to 44 per cent in Rajauri District (Table 8.10). The size of holdings on the whole is small except in the undulating mountainous tracts. The small size of holdings and the fragmented fields have adversely affected the productivity of arable land in most of the districts.

In the state of Jammu and Kashmir land reform measures were initiated, completed and implemented in the early years of 1950. It is, however, unfortunate, that land reforms in the state were restricted only to the redistribution of surplus land among the landless labourers and tenant formers. The other attributes of land reforms such as agricultural extension service, education, supply of modern inputs, development of institutional credits and marketing, etc., were not accompanied by the redistribution of land to the tillers. This has adversely influenced the agricultural production and the economic prosperity of the farmers. The small size of holdings and the non-availability of costly inputs to the small and marginal farmers have constrained innovations diffusion in their holdings as a result of which they, by and large, are the subsistent farmers.

Table 8.10
Jammu and Kashmir: Average Size of Holdings, 1994-95

District	Size and Number of Holdings					
	Below 1	1-5	5-10	10-20	Above 20	Total
Anantnag	94830 (61.96)	56532 (36.94)	1407 (0.92)	184 (0.12)	89 (0.06)	154741 (100)
Pulwama	78135 (79.24)	20909 (13.51)	133 (0.03)	2 (Net.)	7 (0.01)	98604 (100)
Srinagar	82124 (94.24)	4860 (5.58)	99 (0.11)	36 (0.04)	24 (0.03)	87143 (100)
Budgam	87390 (85.61)	14652 (14.36)	35 (0.03)	4 (Net,)	1 (Net.)	102 (82)
Baramulla	115840 (80.95)	26813 (18.74)	398 (0.28)	47 (0.03)	3 (Net.)	143101 (100)
Kupwara	66420 (85.45)	11153 (14.35)	118 (0.15)	16 (0.02)	20 (0.03)	77727 (100)
Leh	6708 (58.81)	4302 (37.72)	322 (2.82)	33 (0.29)	41 (0.36)	11406 (100)
Kargil	11974 (73.11)	4328 (26.43)	69 (1.10)	4 (0.03)	2 (0.01)	16377 (100)
Jammu	94830 (61.96)	56532 (36.94)	1407 (0.92)	184 (0.12)	89 (0.06)	153042 (100)
Kathua	41817 (6015)	25134 (36.16)	2419 (3.49)	134 (0.19)	4 (0.01)	1014489 (100)
Udhampur	55487 (54.67)	43761 (43.11)	2049 (2.03)	176 (0.17)	16 (0.02)	101489 (100)
Doda	63299 (68.04)	29416 (31.62)	305 (0.33)	11 (0.01)	2 (Net.)	93033 (100)
Rajauri	34392 (53.76)	28508 (44.57)	974 (1.52)	89 (0.14)	9 (.01)	63972 (100)
Poonch	29395 (64.91)	15407 (34.03)	449 (0.99)	29 (0.06)	4 (0.01)	45284 (100)
Total	901599 (74.05)	306102 (25.14)	8817 (0.73)	766 (0.06)	225 (0.02)	1217509 (100)

Figures in parentheses show percentages
Source: Digest of Statistics, Jammu and Kashmir, 1994-95

The above analysis of Table 8.10 vividly shows that cereals like rice, wheat and maize dominate the cropping mosaic and in the kharif and rabi seasons and the non-cereals and coarse grains occupy an insignificant hectarage. The predominance of staple food crops in the cropping patterns shows

FIGURE 8.18

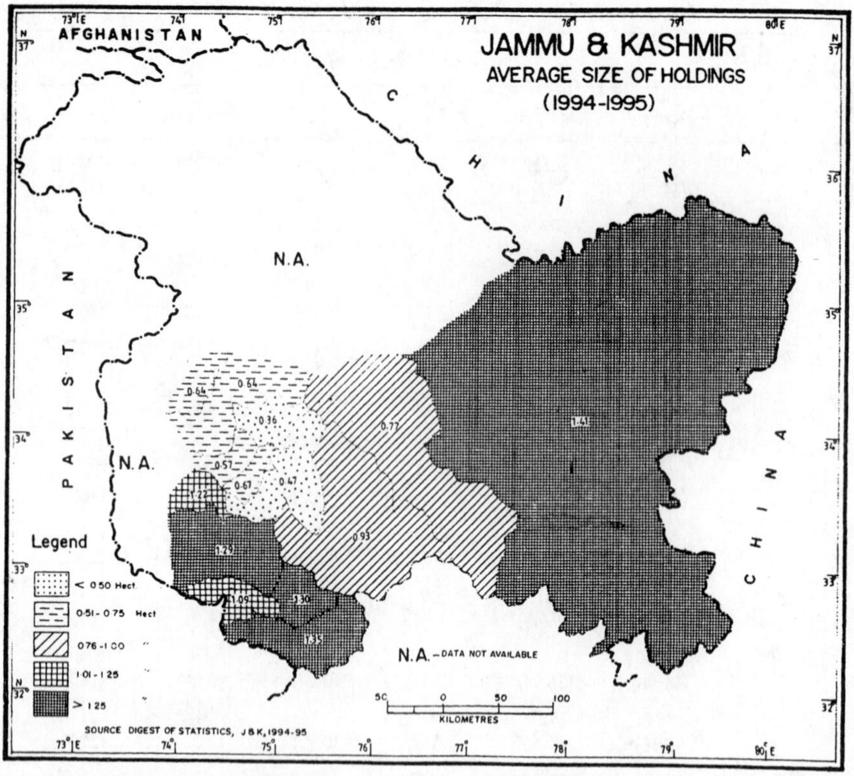

that agriculture in the state is, by and large, subsistent in character and most of the farmers generally grow crops for the family consumption.

Production of Food Grains

Jammu and Kashmir is essentially a grain growing state. Rice, maize, wheat, millets, and pulses are its major food crops. The growth trend of these crops has been given in Table 8.11.

The total production of food grains in 1964-65 which was 9,827 thousand quintals, rose to 13575 thousand quintals in 1994-95, recording an increase of about 38 per cent. The production of wheat has gone up three times and there is 33 per cent increase in the production of rice, while maize recorded an increase of only 16 per cent over the period of last 30 years (Table 8.11).

Table 8.11

Year	*Rice*	*Maize*	*Wheat*	*Other cereals*	*Pulses*	*Total food grains*
1964-65	4327	3718	1122	407	253	9827
1974-75	4560	3036	1931	258	318	10103
1984-85	5690	4551	1684	213	307	12445
1994-95	5775	4325	2995	210	270	13575

Source: Financial Commissioner's Office, Srinagar.

The area under commercial crops in each district of the state has been plotted in Fig. 8.19. It may be seen from Fig.8.19 that in the Valley of Kashmir commercial crops occupy over 20 per cent of the agricultural land except in Badgam in which this is 16 per cent. Pulwama District with 37 per cent, has the highest hectarage under commercial crops. In the Jammu Division 2 to 7 per cent of the arable and is under commercial crops (Fig. 8.19-8.20).

Horticulture

The cultivation of fruits, vegetables and flowers is known as horticulture. The word 'horticulture' has been derived form the Latin word 'Hourtus' meaning garden and 'Colere' meaning cultivate. The concepts of garden culture as a distinct form of field culture dated back to the Medieval farming system that used extensive area of grains, forage, fodder and pastures.

Horticulture is traditionally divided into: (i) pomology, (food crops, nuts, vines, and (ii) floriculture, e.g., edible roots, edible stems, edible leaf, edible seeds and flowers. The state of Jammu and Kashmir has great diversity in the geo-ecological conditions, which suit ideally to the cultivation of almost all kinds of temperate and most of the varieties of tropical and sub-tropical fruits.

Horticulture is an age old economic activity in the state of Jammu and Kashmir. Kalhana, the great Kashmiri historian mentioned fruit culture in Kashmir in his famous book *Rajtarangani* during the reign of King Nara as early as 1000 B.C. It was, however, during the period of Lalitaditya (900 A.D.) that horticulture in the state received considerable patronage. Subsequently, during the periods Zainul-Abidin and the Mughals some varieties of fruits were brought from Kabul, Persia (Iran) and Central Asia.

At present, about 66,000 hectares of land are under orchards in the state. The temporal growth of orchard area has been given in Table 8.12.

FIGURE 8.19

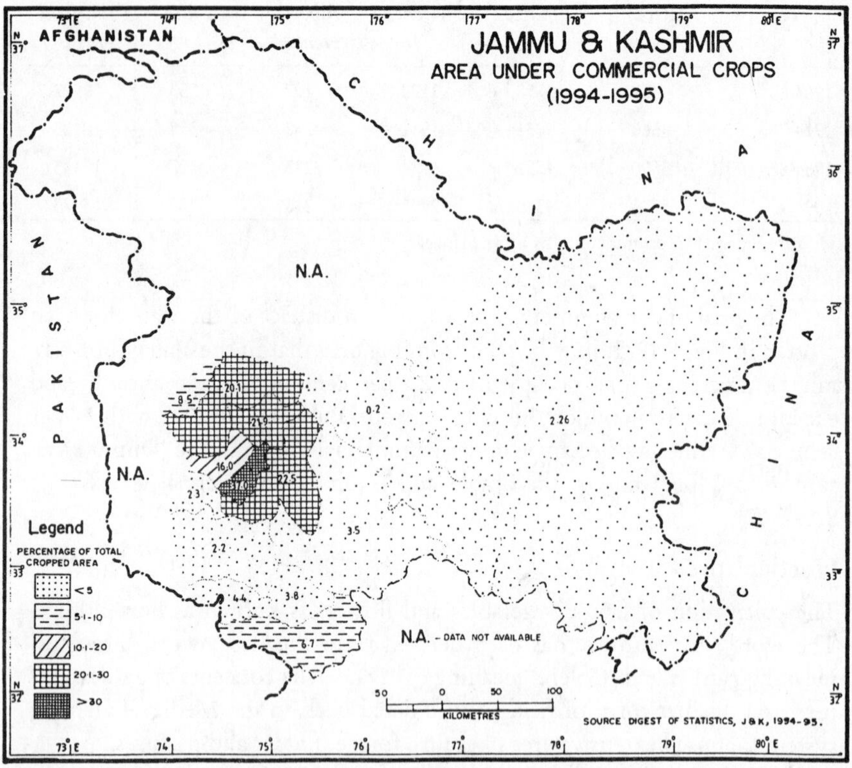

Table 8.12

Jammu and Kashmir: Growth Pattern of Orchards, 1955-95

Year	*Hectares*
1955-56	18,000
1960.61	21,000
1970-71	25,000
1980-81	51,000
1990-91	59,000
1995-96	66,000

Source: Digest of Statistics Jammu and Kashmir, 1995-96.

FIGURE 8.20

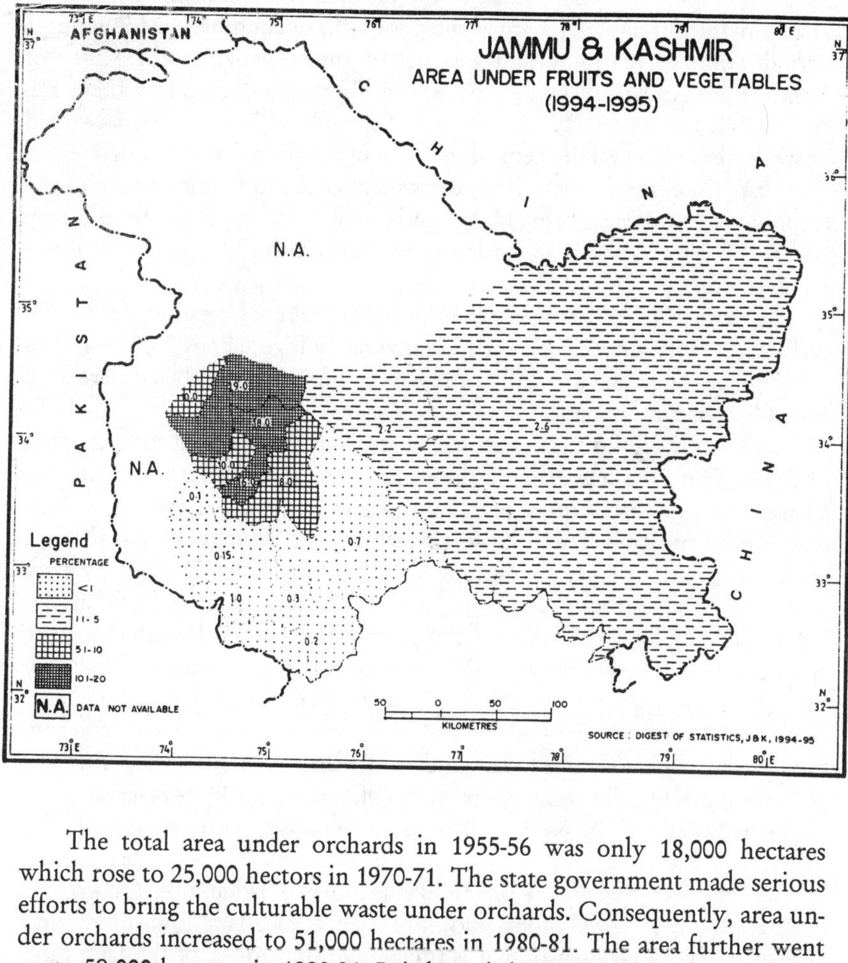

The total area under orchards in 1955-56 was only 18,000 hectares which rose to 25,000 hectors in 1970-71. The state government made serious efforts to bring the culturable waste under orchards. Consequently, area under orchards increased to 51,000 hectares in 1980-81. The area further went up to 59,000 hectares in 1990-91. It is hoped that orchard area will increase to 70,000 by 2001 A.D.

The Valley of Kashmir being in higher latitudes and altitudes, generally grows temperate fruits. It grows numerous varieties of apples, apricots, almonds, peaches, cherries, plums, pears, walnuts, melons, grapes and *Laukats.* The farmers are increasingly concentrating on fruit cultivation and more area is being brought under orchards. It has been reported that several pasture and forest areas have been cleared and planted with fruit trees.

Most of the apple orchards lie at the altitudes of 1,800 to 2,700 m above the sea level. In general, the sandy soil does not suit to apple trees as it induces root-bores and pink disease to the orchard. Deep and heavy loams are more suitable for apple orchard as these soils can retain moisture for a longer period. Some of the important varieties of apples grown in the Valley of Kashmir are Ambri, Delicious, American Teral, Maharaji, Hazratbali, Kesari, and Royal Misri. The districts of Barmulla, Anantnag, Srinagar and Pulwama have over 85 per cent of the total orchards in the state (Fig.8.20).

Pears *(Nakh)* and peach also grow well in cool and humid climate of the Kashmir Valley. Peach, being highly perishable is not much preferred by the farmers. Pears and peach orchards are well scattered in the districts of Kashmir Division.

Apricot *(Khubani)* requires humid nights and cool weather during the early parts of its growing period and moderately high (about 25°C temperature) during later and the hilly areas of Kashmir below 3,000 m are well suited for apricot orchards.

The subtropical fruits are grown in the plains and low altitudinal zones of the Jammu Division. Among the tropical and subtropical fruits, mango, orange, malta, *Kinu, Mausm, Ber,* plum, papaya, guava, *Nashpati* (peach) *Litchi*, lemon, and grapes are noteworthy. Nearly 500 varieties of mango, each with a peculiar taste and flavour, are found in the Jammu Division. Mango takes about four months (March to June) to mature. Mangoes from Jammu region are supplied to Delhi and the towns and cities of Punjab, Haryana and Rajasthan. There were about 3 crore fruit trees in the state in 1995-96. In the recent years, several new projects have been launched to spread orchards in the various parts of the state. Incentives have been given to the growers by the various departments of the state to develop orchards in the drought prone *Kandi* and hilly areas. Subsidy to the extent of 50 per cent of fruit plants and inputs to be used in them are being provided to the growers by the state. Consequently, more than 2400 hectares have been covered under the subsidy schemes. For a steady expansion of horticulture, the grower must apply the required inputs and should remain vigilant against pests and disease. In fact, exact scheduling is imperative since most of the fruits and flower crops are highly perishable. Transportation to market must function smoothly to minimize post-harvest loses.

As per the reported data, area under fruits like apple, walnut, *litche*, pear, cherry and apricot has increased substantially since 1980, while the area of almond, *Lauqat* and peach is shrinking.

On an average, about 40 lakh tonnes of fruits are exported annually. Delhi, Mumbai (Bombay), Chennai (Madras), Hyderabad, Bangalore, Bhopal and the towns cities of Punjab and Haryana are the main markets to which Kashmir export their fresh and dry fruits.

One of the main problems of orchards sustenance is the marketing. The

FIGURE 8.21

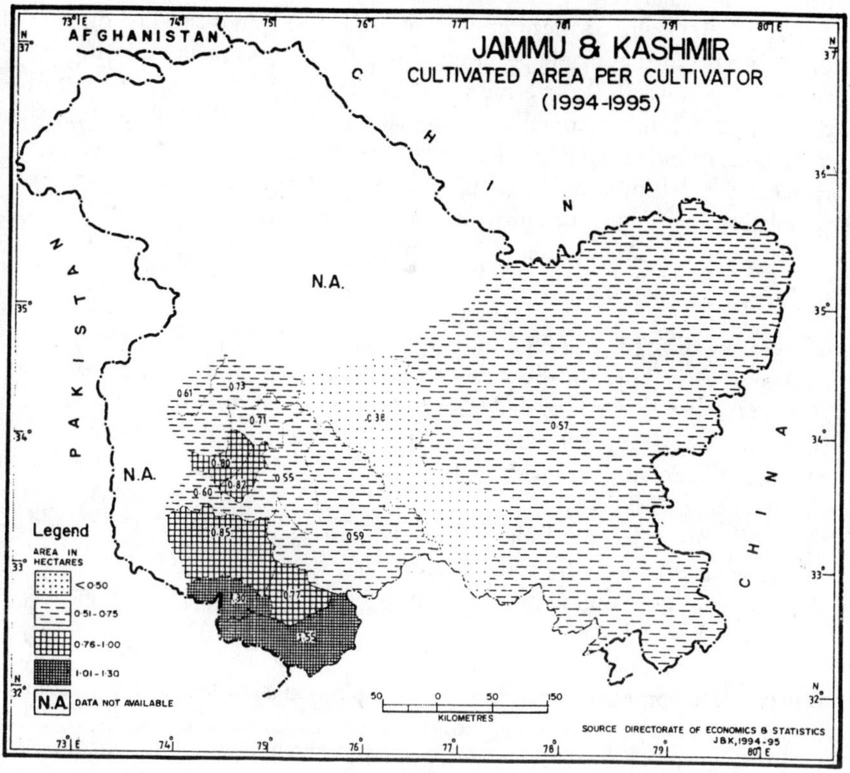

packing and transportation costs are significantly high. Moreover, apple orchard owners are facing a tough competition from the apple growers of Himachal Pradesh and Uttrakhand. The good varieties of apple like *Delicious* are highly vulnerable to pests. The farmers have to spray apple orchards four to six times. The high cost of insecticides and pesticides reduces the margin of profit to the apple growers. There is an urgent need to develop disease resistant varieties, failure to which apple orchards have a less promising future in the Valley of Kashmir.

Land-Man Ratio

The size of holding and fragmentation of fields have a direct bearing on agricultural practices, cropping patterns and yield per unit area. In Jammu and Kashmir like other parts of India, the land-man ratio is very low. The law of succession results in the subdivision and fragmentation of holdings. Consequently, the land-man ratio is declining fast. The cultivated area available per head of population has been plotted in Fig. 8.21.

It may be observed from Fig. 8.21 that the land-man ratio in all the districts of Jammu and Kashmir is below one acre except the Kathua District in which it is 1.55 acre per head of population. In the Kargil and Leh districts the cultivated area available per head of population in 0.38 and 0.57 hectare respectively (Fig. 8.21). The meagre cultivated land available per head of population in Ladakh is mainly because the land is mostly barren and the climate is harsh and arid. Under the increasing pressure of population the land-man ratio is declining at a faster pace. If the present trend in the growth of population continues, the cultivated land available per head of population will be just half to that of the present one after 25 years.

Sericulture

The rearing of silkworms on mulberry trees for the production of raw silk is known as sericulture. In the state, mulberry trees in the wild form existed from antiquity. In Kashmiri language the silkfibre is known as 'pote' and the silkworm is the *Patikeom* (insects). Pote manufacturing in the Valley is an old industry. Sericulture involves the stages of:

(i) Rearing of Silk-worm;
(ii) Collection of cocoons and their delivery for reeling purpose.
(iii) Reeling of raw silk from the cocoons; and
(iv) Availability of mulberry trees from which fresh leaves (the food of silk-worms) can be obtained.

Rearing of silkworm requires scientific knowledge and technical skill. For the development of healthy silk two basic seed stations have been established at Udhampur and Miragund. Work on the Qazigund Nursery is also under progress. There are 2,800 villages and 33,000 households in which sericulture has become an important economic activity. Out of these, 1500 villages are situated in the Kashmir Division and the remaining 1300 villages in the Jammu Division. The area sprawl and production pattern in the state has been given in Table 8.5.

In the state there are about 7 lakh mulberry trees, out of which 53 per cent (370,000) are in Jammu Division and 47 per cent in Kashmir Division. Annually, about 6,680 quintals of raw silk is produced which is approximately worth Rs. 50 crores or 50 million.

Inadequacy of mulberry leaves and damage caused by insects and pests are the major problems in the development and expansion of sericulture. Now the Sericulture Development Department is laying stress on raising dwarf mulberry trees to supplement and replenish the traditional tall mulberry trees. As a result of these efforts, the dwarf varieties have increased substantially in the state. The temporal change in the sericulture development has been given in Table 8.13.

Table 8.13

Jammu and Kashmir: Sericulture Development 1974-94

Year	*Mulberry Trees in 000*	*Production of Raw Silk in 000 kg*	*Value (in lakh)*
1974-75	219.5	70.5	Rs. 176.14
1980-81	601.00	75.85	Rs. 234.98
1990-91	1668.00	20.75	Rs. 155.35
1994-95	1685.00	19.98	Rs. 271.00

Source: Directorate of Sericulture, Jammu and Kashmir Srinagar.

It may be observed from Table 8.13 that in 1974-75 the total number of mulberry trees in the state was 219,500, out of which 82,600 trees were in the Jammu Division and 136,900 trees in the Kashmir Division. The number of mulberry trees in the state increased to 1,668,000 in 1990-91 and recorded a marginal increase on this number in 1994-95, the total number being 1,685,000. There is however, a declining trend in the production of raw silk which was 70.5 thousand kg in 1974-75 and 75.75 thousand kg in 1980-81. In 1994-95, the production came down to only about 20 thousand kg. The substantial decrease in raw silk in the state may be attributed partly to the political instability and partly to the inadequate availability of mulberry leaves to silk worms.

The climate and availability of skilled labour in the Valley of Kashmir are the favourable factors for silk production. New sturdy-hybrid species of silkworm need to be developed which can withstand the adverse climatic conditions. Such species of worms enhance the quantity and quality of raw silk.

Apiculture

Apiculture is also known as bee-keeping. In bee-keeping, bee-colonies are maintained for commercial production of honey and other by-products. Bee-keeping also helps in cross-pollination of crops. Honey produced by honeybees has been considered as a balanced and delicious food for thousand of years. In fact, honeybee is amongst the most ancients of domestic creatures.

The Valley of Kashmir is well known for its great diversity of flowers, ranging from lily to roses and saffron. Most flowers are dependent on bees for carrying pollens from the anthers (male part) of one flower to stigmas (female part) of the flower. Many fruit trees like almond, cherry, peach, pear, pumpkin and walnut do not produce fruits unless bees are present to pollinate them.

Bee-keeping is a household industry in the Valley of Kashmir. There are numerous bee-keeping places in Pampore, Srinagar, Pulwama, Tral, Kulgam, Badgam, Anantnag (Islamabad), Qazigund, Kupwara and Karnah. In Jammu Division bee-keeping is being carried on in Udhampur, Poonch, Kishtwar, Bhadarwah, Batote, Ramban, Doda and Banihal.

Fisheries

The state of Jammu and Kashmir has enormous potential for the development of fisheries. The Valley of Kashmir abounds in numerous water bodies like lakes, ponds, wetlands, springs, streams and rivers. These waters are generally alkaline in nature and are suitable for the culture of local as well as exotic fish. The local fish fauna contains 30 species of which only a few are commercially important. Trout, common crop have been introduced into the valley during the present century. Fish-catching is done by the indigenous technique of cast and dip nets. Long lines are also used in some waters.

Fish, the poorman's protein forms an important item of diet of the local population of Kashmir. There is a continuous demand for the fish from the defence personnels stationed in the valley. The Wular, Manasbal, Dal, Anchar lakes and the Jhelum and its tributaries are the main sources of fish.

Table 8.14

Jammu and Kashmir: Fish Caught in Quintals

Year	*Kashmir Division*	*Jammu Division*	*Total*
1965-66	40480 (94.64%)	2290 (5.36%)	42770
1975-76	65000 (95.30%)	3200 (4.7%)	68200
1985-86	100150 (95.20%)	5050 (4.8%)	105200
1995-96	138000 (98.71%)	1800 (1.29%)	139800

Source: Directorate Fisheries, Jammu and Kashmir.

An examination of Table 8.14 shows that in 1965-66 the total production of fish was 42770 quintals which increased to 1,39,800 quintals in 1995-96. It is interesting to note that the share of Kashmir Division in the total catch in 1995-96 was about 98.71 per cent and the remaining 1.29 per cent is the share of Jammu Division.

The various types of fish found in the valley are Punjabi-gad, Raput-gad, Star-gad, Algad, Chiroo, Kasher-gad, Dap-gad, Chush, Khrong, Guran, Trout, and Anur. Fish culture is being carried on in some of the water bodies of Kashmir. Trout culture is an important economic activity in the valley. It was in the year 1901 that Mitchel procured eyed ova of trout fish from the Great Britain. It was at his initiative that a hatchway was started at

Harwan in a stream. The eyed ova was spread to other streams like the Madhumati, the Dachigam, Nala, and Pohru River.

Since the required conditions for trout culture are very exacting, the number of hatcheries is limited. The main trout hatcheries in Srinagar District are Laribal, and Harwan. Besides the above hatcheries, there is a large trout farm at Achabal in the Anantnag District.

The production from these hatcheries is not sufficient to commercialize the sale of trout. Solman fish from Canada has also been brought and introduced in some of the water bodies of Kashmir.

In brief, there is a great scope for the development of fisheries in the state of Jammu and Kashmir. The local and exotic fish need to be properly diffused in the different water bodies of the state.

Livestock

The state of Jammu and Kashmir has an agrarian economy. There is a close relationship between cultivation of crops and raising of livestock. In fact, both these activities are interdependent and complimentary to each other. The cool climate, the alpine pastures, abundance of fodder and constantly increasing demand of milk and milk products, all favour the keeping of cattle in Jammu and Kashmir state. Cattle, sheep, goats, horses and ponies are kept and reared in large number in Jammu and Kashmir and Leh, is well known for the production of fine wool. The Ladakhi population is largely dependent for their sustenance on livestock, sheep and goats. The number of livestock has been given in Table 8.15.

Table 8.15

Jammu and Kashmir-Livestock 1990 (No. of Animals in lakhs)

	Livestock	*No. in lakh*	*Percentage to Total Livestock*
1.	Cattle (cows)	30.65	38.82
2.	Buffaloes	6.15	7.79
3.	Sheep	25.50	32.30
4.	Goats	15.25	19.32
5.	Horses, ponies, donkies and mules	1.40	1.77
	Total	78.95	100.00

Source: Directorate of Statistics, Jammu and Kashmir, Srinagar.

It may be observed from Table 8.15 that the total number of livestock in the state in 1990 was 78.95 lakh, out of which about 38 per cent were cattle, 32.30 per cent sheep and about 19 per cent goats. Buffaloes and horses, etc.,

FIGURE 8.22

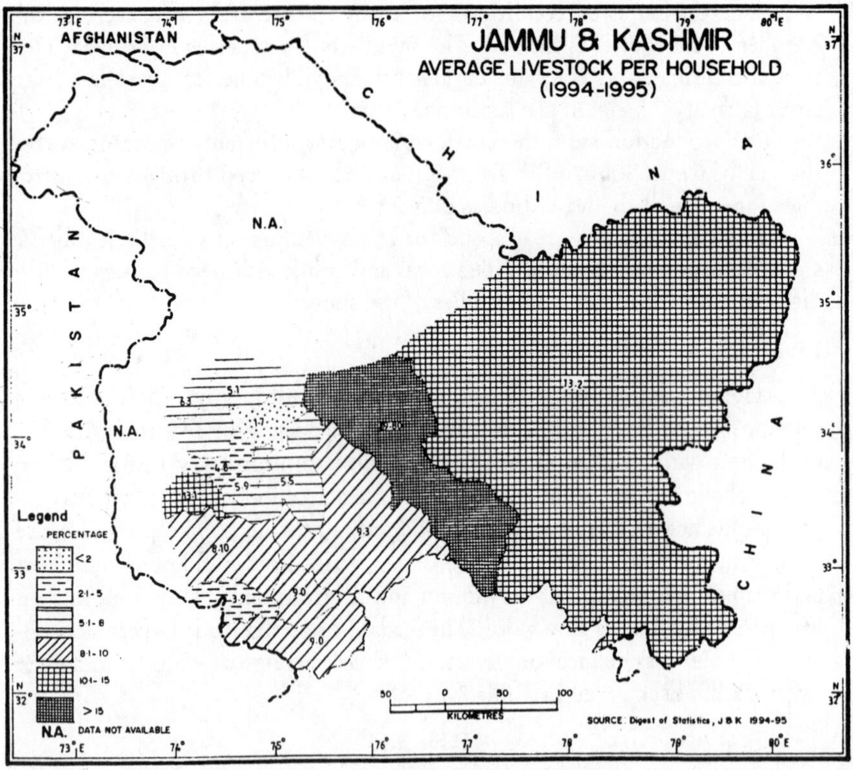

constituted about 8 and 1.77 per cent of the total livestock respectively. The predominance of cattle population is found in the districts of Udhampur, Doda, Kathua, Anantnag, Baramulla, Pulwama and Rajauri.

The average livestock per household in the various districts of the state has been plotted in Fig. 8.21. It may be observed from Fig. 8.20 that the district of Kargil (Ladakh) has the highest number of livestock per household being 20. It is followed by Ladakh and Kupwara in which this ratio is 13 livestock per household. The lowest number of livestock per household is in Srinagar and Jammu, being only about 2 and 4 respectively while in the remaining districts this ratio varies between 4 to 10 (Fig. 8.22). Most of the livestock is however, local and inferior. The state government has been taking a number of steps to improve the quality of cattle, sheep, goats and ponies. A number of cattle breeding centres have been established by the state government. Some of the important cattle breeding centres are:

(i) Cattle Breeding Farm, Chashma-Shahi, and Manasbal, Srinagar.
(ii) Cattle Breeding Farm, Belicharna, Jammu.
(iii) Cattle Breeding Farm, Changspa.
(iv) Cattle cum Karakul Sheep Farm, Khumbathang.
(v) Yak Breeding Farm, Nubra, and
(vi) Veterinary Hospital, Dras.

For the development of better breeds of sheep, the Australian and Russian Marino-sheep breeds were selected for Kashmir, and 16 sheep breeding farms with a total capacity of housing 1900 sheep were established in the different sheep belts. These sheep farms have been located in the areas where grazing facilities are available. In the course of breeding programmes, a new breed of sheep known as Kashmiri Merino has been developed with known body weight (average about 54 kg at the age of two and a half years).

In order to improve the quality of sheep, fine woolled sheep have been imported in piece-meal consignments from abroad. This exotic stock of sheep has been multiplied at sheep breeding farms and the quality *Rams* have been distributed. Their distribution has resulted into cross-breed population of 8 lakh up till 1995. This cross-breed has been examined and it has been found that they have finer quality of wool and increase in mutton as compared to the local population.

Irrigation

Irrigation has been defined as the distribution of water to crops, arable land by artificial means in order to enable or enhance the growth of crops. This irrigation water may be supplied by canals, tubewells, wells, tanks and other sources.

The state of Jammu and Kishmir situated in the subtropical and subtemperate latitudes is characterized with great variations in terrain and altitudes. Consequently, it experiences the extremes in the mean monthly and annual range of temperatures. The distributional pattern and the total amount of precipitation received is also erratic. The scorching heat in the plain of Jammu Division, the uncertainty in the behaviour of summer monsoon in the Valley of Kashmir, and the arid conditions of Ladakh Division compel the farmers to irrigate their kharif and rabi crops frequently to get a good agricultural return. The climate of Ladakh is so arid that cultivation of crops without irrigation is almost impossible.

Irrigation, the process of artificial supply of water to crops, is an age old practice in the state. The farmers are quite conscious of the significance of irrigation as a means to achieve an effective break-through in the development of agriculture. It is reflected in the eagerness with which some of the *Kuls* (canals) were constructed by the cultivators in the past. After independence, several irrigation projects have been envisaged and completed as a result of which the area under irrigation has increased significantly. The cultivated area irrigated in the three divisions of the state has been given in Table 8.16.

Table 8.16

Jammu and Kashmir State: Area and Sources of Irrigation (Area in '000 ha)

Year	*Net Area Irrigated*				
	Canals	*Tanks*	*Tube-Wells*	*Other Sources*	*Total*
1950-51	244	3.00	3.00	11.00	261
1960-61	256	—	5.00	13.00	274
1970-71	254	—	1.00	11.00	266
1980-81	185	2.00	4.00	13.00	304
1994-95	290	3.00	2.00	20.00	315

Source: Financial Commissioners Office, Jammu and Kashmir, Srinagar.

The major sources of irrigation in the state are canals, wells, tanks and others. Canals alone account for about 94 per cent of the total area irrigated in 1994-95, while the remaining about 6 per cent was irrigated by wells, tanks and other sources.

Though the net area irrigated has steadily increased from 261 thousand hectares in 1950-51 to 315 thousand hectares in 1994-95, there is enough scope for development and expansion of irrigation in the Jammu and Kashmir divisions. The net area irrigated in 1994-95 in the state of Jammu and Kashmir has been plotted in Fig. 8.23.

Rice, wheat and maize are the major crops irrigated by canals and tubewells. Jammu District with over one lakh hectares of irrigated land is the most extensively irrigated tract of the state, followed by Anantnag (76 thousand hectares) and Pulwama 56,000 hectares. Badgam, Kathua, and Srinagar are the other districts in which over 25,000 hectares of cultivated land is irrigated. There are striking variations in the net cropped area irrigated in the Kashmir, Jammu and Ladakh divisions. It would be worthwhile to examine the availability of irrigation and sources of irrigation in each of the administrative units of the state.

Irrigation in Kashmir Division

River Jhelum and its tributaries constitute the major drainage system of the Kashmir Valley. Descending from the surrounding mountains, the river of

FIGURE 8.23

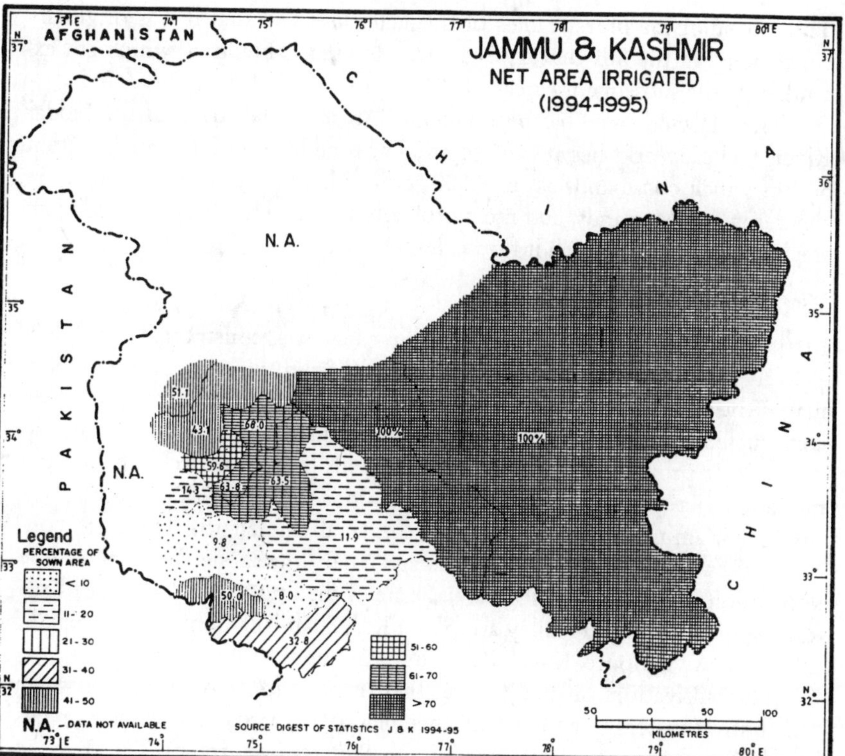

Jhelum and its tributaries flow through the *Karewas* and plain of the valley. The most notable use of water potential of Jhelum in the valley is for irrigation.

The Valley of Kashmir is mainly irrigated by canals. These canals may be divided into:

(i) Zamindari canals.
(ii) Government constructed canals.

The Zamindari canals, locally called as *Kuls* are constructed and maintained by the farmers. The total number of such canals is 2,827 which irrigate nearly 60 per cent of the total irrigated area of Kashmir Division. Out of 2,827 *Kuls*, 259 canals have been taken over by the State Irrigation Department. The major state owned Zamindari canals are:

(i) The Martand Canal
(ii) The Zainagir Canal

The Martand Canal

This is one of the oldest canals of the state. It is said that it was originally constructed in the 8th century A.D. by the king Lalitaditya, which got extended by the subsequent rulers.

The Martand canal has been taken off from the left bank of the Liddar River. Its headworks lies at Ganeshpur. The total length of the canal is about 50 km which commands about 9,500 acres of land. The canal terminates in the karewa of Martands, and most probably its name has been derived from the karewa through which it flows and terminates.

The Zainagir Canal

In the 15th century A.D., the Zainagir canal was constructed by Sultan Zainul-Abidin. This canal has been excavated from the bank of the Madhumati River—a left-hand tributary of the Jhelum. The canal has been constructed along the northern side of the Wular Lake through Sopore and Zainagir. It provides water to the paddy fields and apple orchards of Baramulla District. The total area under its command is about 13,500 acres.

The Zamindari canals are generally small in length. They are constructed in an unplanned manner and their dense network gives the impression of a labyrinth (Fig.8.24). Some of the important Zamindari Kuls (canals) of the Kashmir Valley are: Shara-Kul, Shah-Kul, Sumbal-Kul, Nandi, Zainpura, Awantipore, Kayal, Rishipura and Babul Kuls.

The cultivatiors on the basis of their experience have evolved a water distribution system. At the occurrence of water scarcity, the villagers regulate the supply of water to crops through *Rahviks* (watermen). The Rahviks Committee is formed by the cultivators.

The Zamindari canals, throughout the history, had been of immense help in saving the crops at the occurrence of drought, and they are still a significant source of irrigation in valley. After the abolition of Zamindari System, many of the *Kuls* could not be properly maintained. The ignored and abandoned *Kuls* are proving to be a great menace to their command areas. In winters (Jan. Feb. March) at the occurrence of heavy rains or at the time of heavy melting of snow in June and July, they spill over water and inundate the adjacent agricultural lands, resulting into serious damage to paddy, oilseeds, maize and saffron crops. In 1980s, a comprehensive plan was prepared for the restoration and maintenance of Zamindari canals. Efforts need to be made to extend these canals in the dry farming areas, especially in the *Wudur* tracts (karewa lands). By providing water to the unirrigated undulating karewas, more area may be brought under cultivation which may be used for the development of orchards and cultivation of paddy, maize, oilseeds and vegetables.

FIGURE 8.24

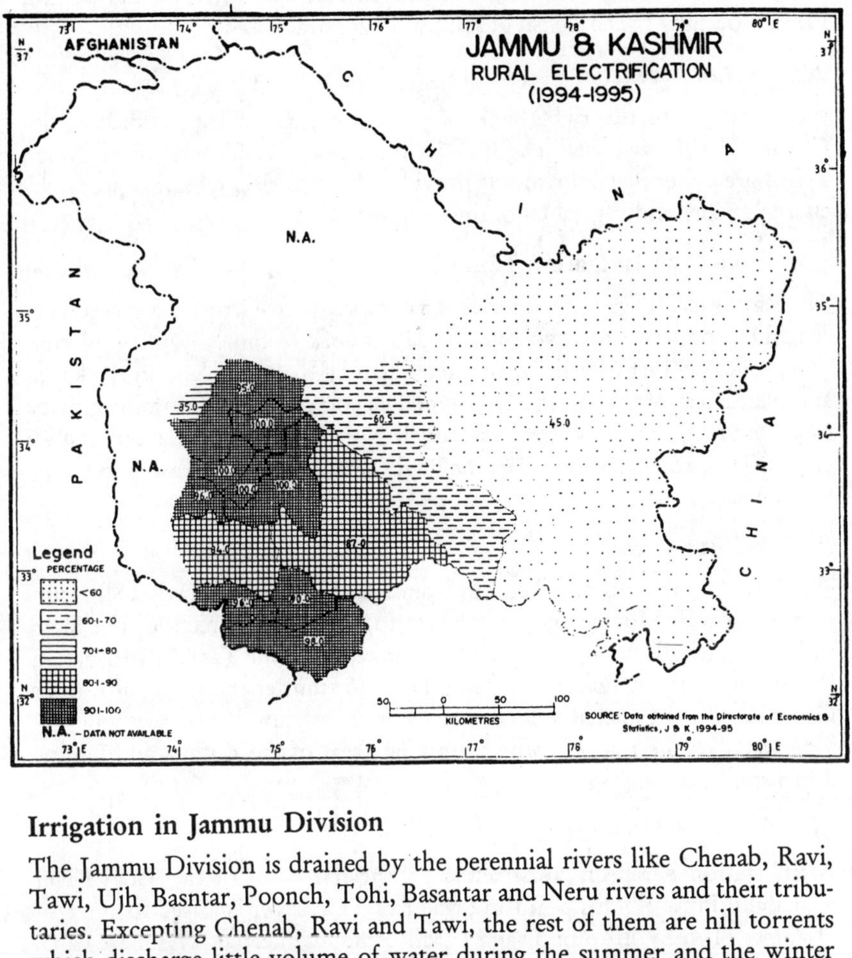

Irrigation in Jammu Division

The Jammu Division is drained by the perennial rivers like Chenab, Ravi, Tawi, Ujh, Basntar, Poonch, Tohi, Basantar and Neru rivers and their tributaries. Excepting Chenab, Ravi and Tawi, the rest of them are hill torrents which discharge little volume of water during the summer and the winter seasons. Wherever possible, the water of these rivers has been utilized for irrigation and hydel power generation purposes. Some of the important canals of Jammu Division have been discussed in the following paragraphs.

Ranbir Singh Canal

The Ranbir Canal has been taken out of the left bank of Chenab and its headworks lies at Akhnoor. Passing southwards, it has been taken across the Tawi River with the help of an aquaduct. The productive plains of Jammu and Ranbir Singhpura are under its command. Its canal and command area is over one lakh acres, this canal and its distribution have helped appreciably in the adoption of High Yielding Varieties of rice and wheat in the Jammu Plain, especially in the Tehsil of Ranbir Singhpura.

The Pratap Canal

Taken out from the right bank of the mighty Chenab at Akhtoor, the Chenab Canal has a total length of 36 km. About 20,000 acres of Akhnoor Tehsil are under its command. It provides water to rice, wheat, oilseeds, orchards, fodder and vegetables of the command area.

The Ravi-Tawi Irrigation Complex

The Ravi-Tawi Irrigation Complex is a comprehensive irrigation project, designed to connect these two rivers by a network of canals. On completion this project will irrigate about one and a half lakh acres of land in the kandi and plain areas of the Jammu Division. The complex is based on the gravity canals and lift irrigation projects which is expected to cost about Rs. 70 crores. The project consists of (i) the Ravi Canal, and (ii) the Tawi Lift Irrigation Project.

(i) The Ravi Canal

The Ravi Canal has been taken out from the right bank of Ravi at Shahpur-Kandi Barrage. Moving in a north-east to south-west direction, the main canal has a length of 76 km. It has been linked with the Tawi Lift Irrigation Project at Vijaipur. Efforts are being made to transfer the water of Kathua Canal into Ravi Canal at Lakhanpur. After completion the project will provide water to the rice and wheat growing areas of the Kathua and Jammu Tehsils.

(ii) The Tawi Lift Irrigation Project

A lift station has been constructed on the left bank of the Tawi River near Bahu Fort. At this point the water of Tawi river is raised to a height of about 31 m by lift pumps and fed into the main canal. The main canal has a length of about 29 km which has been constructed from Jammu to Ramgarh. It commands an area of about 15,000 acres, stretching over the blocks of Bishna, Jammu and Vijaipur.

Ladakh Division

The desolate terrain and scanty rains in Ladakh are not conducive for the cultivation of crops unless irrigation to the field is available. In fact, the entire cultivated land of Ladakh is irrigated. Some of the growers have dug their personal irrigation channels to irrigate their fields. In addition, there are several small canals, locally known as *Mayur*. These canals (Mayurs) play a vital role in the agricultural landuse of Ladakh.

(i) The Chuchot Mayur

The Chuchot Mayur has been taken out of the left bank of the Indus at Stakna. It commands about 200 acres of land in the villages of Chuchut, Gongma, Shama, Yokma and Balam. Wheat, *Grim* (a type of barley) vegetables and fodder are the main crops irrigated by this canal.

(ii) The Thikasay Mayur

This Mayur (canal) has been excavated from the right bank of the Indus River. It provides water to the villages of Ranbirpura, situated on a low terrace of the Indus. About 120 acres of cultivated area of Ranbirpura village is under the command of the Thikasay Canal.

Tank Irrigation

Tank irrigation is not very common is the state. It is a source of irrigation in some of the districts of the Valley of Kashmir. About 3,200 hectares of arable land are irrigated by tanks out of which 97 per cent are in the Anantnag, Pulwama, Srinagar, Badgam and Kupwara districts and only about 3 per cent in the Poonch District.

Hydel Power and Energy Resources

Energy resources hold the key to the economic development of a geographic region. Their availability accelerates growth and development, while their non-availability retards the process of development. The level of consumption of energy in any country is a significant indicator of development. For instance, the U.S.A. claims a per capita consumption of about 9,000 units per day. In sharp contrast, the Indian average per capita consumption is only 130 units. In Jammu and Kashmir, it is only 90 units per head per day. Thus, it is one of the most backward states of India and may continue to stay as such unless adequate and early steps are taken to improve its power situation. In the following paragraphs an attempt has been made to present a succinct picture of the energy resources of the state of Jammu and Kashmir.

The state of Jammu and Kashmir though poor in coal, petroleum and natural gas, has tremendous potential for hydel power generation which has been inadequately tapped.

The three perennial rivers—Indus, Jhelum, and Chenab and their tributaries which drain the state, open immense opportunities for the generation of hydroelectricity. It has been estimated that these rivers provide theoretically a potential of about 13,000 MW of hydel generation. Because of faulty planning and priorities, only a fraction of this potential has so far been utilized (Table 8.17).

Table 8.17

Jammu and Kashmir State: Generation of Hydel Power (1984-85)

Division		Installed	Effective Capacity MW		Summer Winter Difference
			Summer	Winter	
A. Jammu					
(1)	Chenei HEP	25	20	12	8.0
(2)	Kalakot Thermal	21.4	4	4	—
(3)	Other Small Stations	3.0	2	1	1.0
	Total A	49.9	26	17	5.0
B. Kashmir					
(1)	L. Jhelum J.P.	105	105	65	40.0
(2)	U. Jhelum Stage I	22.62	22	17	5.0
(3)	Ganderbal	15.0	12	10.5	1.5
(4)	Morhra	9.0	7	7	—
(5)	Bemina and others	6.0	—	4.5	—
	Total B	157.6	146.0	104.0	—
	Total A + B	208.1	168.0	121.0	47.0
C. Imports					
	State share from B.B.N.M.	—	27.0	27.0	—
	Total A + B + C	208.1	195.0	148.0	47.0

Source: Digest of Statistics, Jammu and Kashmir State, 1993-94.

The total hydel power generation in the state amounts to 168 MW in the summer season, while it drops to 121 MW in the the winter season. In addition, the state gets 69 MW from the northern Grid which, with the help of capacitator, is raised to 75 MW. These sources of power together supply an average 956 MU of power per year which is almost half the actual demand for power in the state.

Lack of adequate power supply has retarded the growth of industrial, commercial and agricultural development in the state. Power supply is rationed throughout the year and particularly during the winter season, when in Srinagar (capital) no part of the city gets more than 10 hours of power supply in 24 hours. On account of this restricted supply of power in the state, and especially in the valley, no major industry, except the Hindustan Machine Tools (HMT) could so far be attracted despite packages of incentives, such as tax concessions and rent-free land, which the state government has offered to provide. If this gap in the demand and supply is allowed to grow, the state will perpetually remain one of the least developed states of India.

Hydel Power Projects under Investigation and Execution

Although, theoretically the potential for power development seems to be colossal, a realistic estimate of the potential for power development reveals that by the end of present century, if financial resources are made available and political stability restored, about 3200 MW hydel power may be installed in the state. The expected installed capacity of the three regions of the state has been given in Table.

Table 8.18

Hydel Projects Under Investigation

	Region	*Installed Capacity*	*Energy Million Units*
A.	*Ladakh Region*		
1.	Iqbal Rridge Hydel Project	3.75 MW under execution	
2.	Igo-phey	5.60 MW	38.65
3.	Nattayan	0.45 MW	0.25
4.	Shakerchiktan	0.26 MW	1.56
5.	Timbis Kunur	0.20 MW	1.20
6.	Haftal	1.00 MW	2.17
7.	Kumdock	0.10 MW	0.60
8.	Dum Khar (micro)	2.00 MW	2.20
9.	Tungate	.10 MW	0.60
10.	Bung-Dong	.20 MW	1.20
11.	Thusam barso	.20 MW	1.20
12.	Marpoch	.20 MW	1.20
13.	Hanu	.75 MW	1.50
		11.06 MW	48.65
Medium Hydel Projects			
1.	Panikher Parkchak	9.5 MW	21.19
2.	Chotak Project	9.5 MW	21.19
3.	Dumkhar	9.5 MW	21.19
	Total	28.5 MW	63.57
	Grand Total	39.56 MW	111.71
B.	*Kashmir Region*	*River Stream*	*Installed Capacity*
1.	Upper Sindh Hydel Project Stage II	Sindh	10.5 MW
2.	Nunwan-Batkote	Liddar	22.6 MW
3.	Shitkari-Kulan	Sindh	84.0 MW

Contd...

Contd...

4.	Boniyar	Boniyar	4.8 MW
5.	Hirpora	Rambiara	9.0 MW
6.	Athwatoo	Madhumati	7.5 MW
7.	Barenwar	Doodganaga	2.0 MW
8.	Tangmarg	Ferozpur Nallah	2.6 MW
9.	Mutchill		1.0 MW
10.	Asthan	Gurez Nallah	0.6 MW
11.	Keran	Keran Nallah	2.6
12.	Wangat	Wangat	1.5 MW
13.	Ardow Stage I	Liddar	7.5 MW
14.	Ardow Stage II	Liddar	3.0 MW
15.	Laripora	Liddar	2.5 MW
16.	Erin	Erin	3.0 MW
17.	Pahalgam	Liddar	3.0 MW
	Total		239.8 MW
C.	*Jammu Region*	*Installed Capacity*	*Location/Investigation Agencies*
1.	Gyspa	222 MW	NP/HP
2.	Bardung	54 MW	NP/HP
3.	Seli	87 MW	NP/HP
4.	Roli	295 MW	NP/HP
5.	Kirtha Naunatoo	750 MW	NP/HP
6.	Naunatoo-naigad	400 MW	J&K/J&K
7.	Dul-Hasti	390 MW	J&K/CWC
8.	Ratlee	170 MW	J&K/CWC
9.	Bagliar	220 MW	J&K/CWC
10.	Sawel Kot	400 MW	J&K/CWC
11.	Bursar	275 MW	J&K/CWC
12.	Pukhul	325 MW	J&K/CWC
13.	Salal	354 MW	J&K/CWC under execution

Source: Digest of Statics, Jammu and Kashmir.

It may be noticed from Table 8.18 that most of the projects which have been taken up for investigation and execution in the Kashmir and Ladakh divisions are either medium or mini-order projects. Such projects can be completed within the stipulated time, if funds are made available. In contradiction, most of the projects in Jammu Division are large-sized and highly capital intensive. The power development programmes in the early stages of the plan periods were installed because of the emphasis of state planners on major hydel projects, such as Salal and Dulhasti, which if completed would have yielded over 500 MW of power. Unfortunately, the planning Commission could not allocate requisite resources for these projects.

In order to accelerate the process of development and to remove backwardness, resources must be found to implement these power projects which are either in advanced stages of investigation or execution.

Power Demand and Supply Projections

The state of Jammu and Kashmir has a weak statistical base. Time series data are not available for scientific projection of power demands for the next 20 years by the various sectors of economy. The unstable political conditions of the state especially that of the Valley of Kashmir are the main barriers in making reliable projections. Even the present consumption figures for the various sectors are not available and have been disaggregated in accordance with the growth estimate of power demand projected by the state government. These points may be borne in mind while having a glance at reading Table 8.19, giving existing and projected power consumption in the state by 2000 A.D.

Table 8.19

Present and Projected Consumption of Power by Sector-1984-88, 1989-90, 1990-95, and 1995-2000 A.D. Million Units

Sector	*Power Consumption 1989-90*	*Projected Consumption Power Increase*			*Projected Rate in 2000 AD over 1984-5 level of Consumption*	*Growth in per cent*
		1989-90	*1990-95*	*1995-2000*		
Total	956.0 MU				4834.14	
Agriculture	122.0	172.5	266.4	436.0	134.0	10-15
Industry	250.0	502.8	1015.6	2392.8	2142.8	12-20
Commerce	80.0	150.6	379.3	770.2	690.2	10-11
Sewage Public Water Supply and Municipal Corporation	90.0	137.5	242.3	478.3	388.3	12
Public Lighting	12.0	23.49	41.98	89.04	72.04	12
Bulk Consumers, Commercial Non-Commercial	180.0	262.0	421.90	745.50	665.50	10
Transport	—	61.03	235.30	57.32	571.32	—
Domestic	734.0					
Estimates	222.0					

Source: Digest of Statistics, Jammu and Kashmir.

FIGURE 8.25

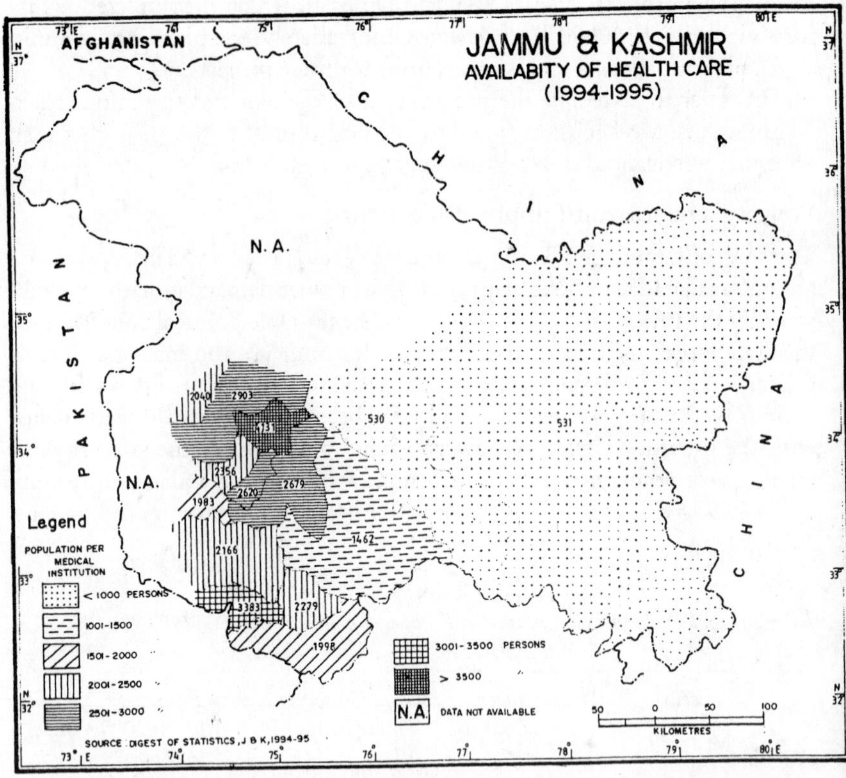

The energy planners of the state have projected a five-fold increase in the consumption of power, excluding domestic consumption from 956 MU in 1984 to nearly 5000 MU in 2000 A.D. The highest rise in demand has been projected in the industrial sector on the assumption that a high rate of industrial growth is inevitable to cope with the rising unemployment. Industrial growth will be dependent on the availability of power and the political stability. The domestic sector is also likely to register phenomenal growth if other sectors develop along the lines projected above. The projection figures make it declare that economic well-being and future prosperity are inextricably tied with massive power development.

Rural Electrification

Rural electrification is a pre-requisite for an integrated and sustainable development. The lifestyle in the rural areas goes under radical transformation once electricity is provided. Being hilly and mountainous and drained by perennial rivers, the state is favourably placed from the point of view of hydel power. Unfortunately, the full potential of hydel power has not been harnessed in the state. Consequently, there is an inadequate supply of power in the months of December, January, and February, when the winters are snowy and severe. The picture of rural electrification has been presented in Fig. 8.24.

It may be observed from Fig. 8.24 that almost all the villages of Kashmir Division are electrified except Kupwara in which only 85 per cent villages are electrified and Baramulla in which 95 per cent of the villages have electricity. Over 90 per cent of the villages of Kathua, Jammu and Udhampur have electricity, while in Doda and Rajauri 87 and 84 per cent of the villages are electrified respectively. Most of the villages of Leh District and about 40 per cent villages in Kargil do not have electricity supply (Fig. 8.24). The state government is making sincere efforts to provide electricity to each and every village, and it is hoped that by 2010 A.D. the entire rural settlement of state will have electricity supply.

Health Care

The average life in the state of Jammu and Kashmir is shorter as compared to the adjoining states of Punjab and Himachal Pradesh. It may be partly because of the physiological and environmental factors and partly because of the inadequacy of health care centres and family welfare schemes. The district-wise distribution of health care facilities has been shown in Fig. 8.25.

An examination of Fig. 8.25 shows that in general one medical institution is available to 2,000 of population. In the state, both medical and paramedical manpower fall grossly short of the officially recommended minimum norm. In the districts of Srinagar and Jammu, there is heavy pressure on medical institutions. The hospitals and medical institutions of the Jammu and Srinagar cities are under great stress which deserve urgent attention of the government. The last decade witnessed a stagnation in the hospital beds-population ratio. There are startling differences between rural and urban areas, pointing to the large urban bias in the provision of health-care infrastructure.

Notes

1. Gathercoal, E.N., E.H. Wirth, 1947, Pharmacognosy lea and Febigor, Philadelphia, p. 172.
2. Stein, M.A., 1979, Kalhan's Rajtarangni: A Chronicle of the Kings of Kashmir, First Trang, P. 13
3. Fazal, Abul, 1876, *Ain-i-Akbari* (ed.), Blochmann, Calcutta.
4. Munshi, A.M., Sindhu, J.S. et. al., 1989, Improved Cultivation Practices for Saffron, *Indian Farming,* June p. 27.
5. Husain, M., 1997, *Systematic Agricultural Geography,* Rawat Publications, Jaipur.

9

Industries

Industries fall under the secondary economic activity. The industrial process involves changing the form of goods to enhance their value. To undertake the manufacture of goods, inputs in the form of capital, labour, power and raw materials are required. The output is a finished product that can either be used again as a raw material for another manufactured or consumed in its present form. The location of an industry, thus, largely depends on the availability of raw materials, minerals, power, capital, labour, infrastructure and managerial skill. The establishment of industry is also influenced by the general climatic conditions, weather, industrial inertia, historical accident and the government policy.

The state of Jammu and Kashmir, though rich in water and forest resources, has very few metallic mineral resources. The non-availability of iron-ore, copper, good quality coal, petroleum and natural gas are the major constraints in the development of basic industries and manufacturing centres. Nevertheless, the Kashmiris have an age-old tradition in the manufacturing of carpets, silk textiles, shawls, raffle, woodwork and handicrafts. In the rural areas, leather industry, oil-crushing, pottery, blacksmithy, carpentry, paper machie, willow-wicker, soap making, food processing, cricket bat and toys making are some of the important industries which provide full or part-time employment to the people.

The Kashmiris have won a great reputation as artisans and were celebrated in the old days for their skill in art manufacturing. The chief centre of Kashmiri industries is Srinagar, but other localities are also famous for their special manufactures. For example, Islamabad (Anantnag) turns out excellent embroideries, Kulgam is famous for lacquered woodwork. Bijbahera has a reputation for its excellence in wood-carving, while the villagers of the Zainagir Circle are famed for their soft woolen cloth. Every Kashmiri seems to excel as a weaver, and the home-spun cloth woven by the villagers in the winter is highly appreciated all over the world. Some of the important industries of the Jammu and Kashmir state have been concisely described in the following paragraphs.

Silk Textile

Silk textile is one of the most ancient industries of Jammu and Kashmir state. Kashmiri silk-goods are renowned the world over for their quality, colour and shades. There are historical evidences which prove that silk fabrics were used to be exported to Persian, Greeks and Roman empires. During the medieval period, the Mughals were the great lovers of silken clothes. They patronized this industry in the Valley of Kashmir.

According to the data of 1995-96, silk industry and its allied activities provide employment to about 2.50 lakh people and,contribute about Rs. six crores (60 million) to the income of the Jammu and Kashmir state. It also provides raw material for shawl making, carpet, *gabba*, *namda*, hosiery and embroidery making. Moreover, it helps in the utilization of culturable waste and less productive tracts for the various activities of silk textile. The number of persons engaged in the various activities of silk has been given in Table 9.1.

Table 9.1

Jammu and Kashmir: Employment in Silk Industry, 1994-95

	Employment	*Number*	*Percentage to Total Workers*
1.	Silk factories	1150	3.70
2.	Weaving factories	2100	6.77
3.	Mulberry culture	550	1.77
4.	Staff seasonal workers	1000	3.23
5.	Grainage	1200	3.88
6.	Silkworm rearing	250,00	80.65
	Total	31000	100.00

Source: Digest of Statistics, Govt. of Jammu and Kashmir, 1994-95.

It may be observed from Table 9.1 that silk textile provides employment to about 3.1 lakh workers. About 85 per cent of the workers are busy in silkworm rearing and grainage. There are 1,150 skilled and unskilled workers who are employed permanently in silk textile.

The modern silk factory consists of several sectors. The basic units comprise of: (i) farmers, rearing silkworms and production of cocoons, (ii) the second unit is the manufacturing of silk fibre and cloth. There are two silk factories in the state. One of them is located at Jammu. The Department of Sericulture Development which produces improved varieties of silkworms takes care of the mulberry trees.

The Rambagh Silk Factory was established in May 1897 by Raja Ranbir Singh under the supervision of Malton. The annual production data of the Rambagh Silk Factory have been given in Table 9.2.

Table 9.2

Rambagh Silk Factory, Srinagar-Silk Production, 1994-95

Year	Production of Raw Silk in kg
1974-75	50,792
1980-81	57,850
1985-86	43,520
1990-91	42,125

Source: Data obtained from the Silk Factory, Rambagh, Srinagar.

The Rambagh Silk Factory produces about 50 thousand kg of raw silk annually. The silk production, however, fluctuates between 37,361 kg in 1982-83 to 57, 850 kg in 1991-92. The production of silk products largely depends on the availability of mulberry leaves, the main food of silk-worms.

There are various categories of workers in the Rambagh factory who perform different functions. The number and proportion of workers who are engaged in various activities of raw silk production in the Rambagh Silk Factory have been given in Table 9.3.

Table 9.3

Rambagh Silk Factory: Categories and Number of Workers-1995

	Category of Workers	No	Percentage to the Total Workers
1.	Spinners	325	33.68
2.	Cooks	175	18.13
3.	Knotters	40	4.15
4.	Silk cleaners	60	6.22
5.	Waste cleaners	55	5.70
6.	Rewinders	50	5.18
7.	Store keepers	225	23.32
8.	Others	35	3.62
	Total	965	100.00

Source: Data Obtained from the Silk Factory, Rambagh, Srinagar, 1995.

It may be observed from the Table 9.3 that about one-third of the total workers (33.68%) are the spinners, about 23 per cent are storekeepers and about 18 per cent are the cooks. The remaining about 25 per cent workers are the cleaners, knotters and rewinders (Table 9.3).

Agriculture in Jammu and Kashmir state, despite favourable geo-climatic conditions is not developing at the desired level. Apart from the vagaries of weather, competition from artificial silk is posing serious problems for silk industry as the synthetic fibre is cheaper to that of pure silk fibre. Other problems of the industry include, low productivity, and the inferior quality of cocoons. High cost of production, absence of marketing agency and the fluctuating demand for silk are the other serious problems of the industry.

Carpet-Making and Woolen Textile

Carpet-making is one of the oldest industries in Kashmir. Kashmiri carpets are famous all over the world for their excellent designs and natural patterns. Though carpets are made in almost all the towns of the valley, their major factories are in and around the City of Srinagar.

In the manufacturing of Kashmiri carpets, the warp is drawn in cotton, while the leaves and texture, leaving a fluppy pile is done by wool, silk and synthetic fibres. The number of knots per sq cm/inch determines the quality and value of carpet, together with the quality of yarn, dye-stuff and finish. Kashmiri *qaleens* (carpets) are manufactured by the government undertakings as well as by the private manufacturers. Some of the important carpet manufacturing centres in Srinagar are, the Cottage Industry Exposition, C.A.E. Carpet Factory, the Kashmiri Carpet Factory, the East-India Carpet Factory, the Oriental Carpet Factory and the John Carpet Factory.

In most of these factories, children and teenagers from the poor families are employed. These workers get low wages. Having inadequate nourishment, they work under unhygienic conditions. Consequently, their health, efficiency, literacy and education are adversely affected. About 75 per cent of the total carpets production is exported to the countries of Middle East and North-West Europe (U.K., France, Netherland, Germany, Denmark, Italy and Belgium). Carpet export is one of the leading items of foreign exchange earners.

Carpets-making has many allied and ancillary crafts and cottage industries. *Namda* and *Gobba* are the special types of woolen carpets, generally used by the Kashmiris to combat cold. *Namda* is a type of felt made of raw wool and cotton mixed in different proportions according to their grade and quality. The average size of *Namda* is 1 m × 1.5 m. Srinagar City is the main centre of *Namda* manufacturing. *Gabba* is relatively large in size, being about 1.75 m × 2.50 m in dimension. *Gabba* is made of coarse wool in black, red, yellow and green colours combinations. The *Gabba* of Islamabad

(Anantnag) are famous all over the country and abroad.

Apart from carpet-making, there are two woolen textile factories, one each at Karan Nagar and Bemina in the City of Srinagar. The wool textile factory of Naushahra (Srinagar) procures fine quality of raffle, utilized mainly for shawl making. Woolen hosiery goods are produced in the Jammu City. About 900 workers are engaged in the woolen textile industry in the state, fetching about Rs. 3 crores (Rs. 30 million) annually.

In 1994-95 about 82 thousand people were employed in the carpet-making and about 10 thousand in *Namda* and *Gabba* making. Thus, carpet-making is an important employment generating industry. The industry is however, facing a number of problems as the developed countries (Canada, U.S.A., Germany, France, U.K. etc.) have banned the import of Indian carpets produced by children. Apparently, the objective seems to stop the exploitation of child labour. This policy of the developed countries is however, coming in the way of growth and development of carpet industry in the state of Jammu and Kashmir.

Forest-based Industries

The state of Jammu and Kashmir has about one-third of its total area under forest. Most of the forest species in the higher altitudes belong to the conifers, while in the lower altitudes pine and deciduous broad-leaves trees are more prominent. These forests provide raw material to a number of forest-based industries. Paper, pulp, match, delicate boxes, sports goods (cricket bats), furniture, joinery, toys, artifacts and decoration pieces are some of the agro-based industries well developed in the Valley of Kashmir.

Although a number of joinery mills have been established in Srinagar, especially along the Srinagar-Baramulla Road, Pampore, and Jammu, sports goods are being manufactural at Miran-Sahib (Jammu) and in the villages of Anantnag District. There is an urgent need to develop forest-based industries in the state on a scientific basis. The willow, mulberry and walnut trees can provide raw materials required for the development of sports goods, furniture, and wood-artifacts. Nearly 5,000 workers earn their livelihood from the forest-based industries and their annual production amounts to more than Rs. 5 crores (Rs. 50 million).

Forest also provide turpentine and a variety of resins, used is several chemical industries. Resin is collected from the pine trees of Jammu and Kashmir Division. It has diversified properties. Lac obtained from the forests can be utilized for the manufacture of polish, gramophone-records, adhesive, printing ink, etc. The resin processing and manufacturing centres are situated at Miran Sahib, Sunderbani, and Rajauri. These three factories employ about 450 workers and produce resin worth Rs. 10 lakh.

The carpenters of the villages are extremely adroit and do excellent work. Some of the *lattice-work* and carving of the shrines is very beautiful, and argues a strong artistic instinct. The skill of the carpenter is the more to be admired when one considers the primitive and indigenous tools with which he works. With a kind of small hammer, half adze (*Tur*), and chisel (*Turats*), the rural carpenter executes any work which his client may require. Allied to the carpenter are the axemen and sawyers. The Kashmiri prefer axe-cut timber both for houses and boasts, and a boat made of axe-cut timber fetches more than one made of sawn timber.

The basket industry is also of importance. Most of the villages have their artisans who make the necessary basket, and baskets for agricultural purposes, and the *Kiltas* used for the transport of apples and for rough village work.

The Kashmiri carvers are well-known for their skill in woodwork. The skilled carpenters prepare beautiful ceilings with perfect designs which are cheap and effective as well. This type of ceiling is known as *Khatamband*. The shrine of *Naqshbandi* and that of the tourists reception centre are some of the excellent examples of Kashmiri woodwork.

The boat industry of Kashmir is also of great importance. The Hanz or Hanji, as the boatmen are called, are the boat-dwellers in the various water odies. The boat industry is quite old in Kashmir. Now forest conservancy and reduction in forests area have made good quality *Deodar* wood very expensive which is coming in way of development of boat industry.

Agro-based Industries

The state of Jammu and Kashmir has an agrarian economy. In fact, agricultural products not only yield over 50 per cent of the states Gross Domestic Product (GDP), it provides raw materials to a number of industries. Fruit-canning, edible oil extraction, flour mills, rice-husking factories, bakery and alcohol preparation draw their raw materials from agriculture.

The plain areas of the Jammu Division and the Valley of Kashmir produce huge quantities of rice. Over 60 per cent of the total population of the state is rice eater. Consequently, there are numerous rice-husking factories in the state, situated mainly in smaller towns of the rice growing areas. The rice mill of Barbarshah (Srinagar) is quite large. A modern rice factory was established at Laithpora (near Pampore) in 1981. The rice husk and rice bran are used for the extraction of fatty oil which finds application in soap-making industry.

The Valley of Kashmir has large tracts under apples, almond, walnut, cherry, peach and pear orchards. Transportation of these perishable fruits to the distant markets by roads is quite expensive. The processing of fruits, making jam, jelly, juice, etc., is an important industry in the state. Numerous fruit processing and canning factories have been located in Baramulla and Anantnag districts. The Food Corporation of India (FCI) has to take initiative towards the establishment of more food and fruit processing factories.

Papier Mache

Papier mache is made from the pulp of paper. The lacquer-workers apply their beautiful designs to smooth wood. These designs are very intricate, and the drawing is all freehand. The pen-boxes (*qalamdan*), tables, cabinet, trays, boxes are the main articles of papier mache. Papier mache still has great national and international market. After 1989, papier mache suffered as the disturbed political conditions discouraged the arrival of tourists. Papier mache industry is largely confined to the City of Srinagar and its adjacent areas.

Kashmir is well known for the production of leather goods. In many of the villages around Srinagar, Islamabad (Anantnag), Baramulla and Badgam, hides are prepared by the *Watals* and then are sold in the markets of Srinagar where they undergo a refining process. Skins are brought in raw, and are prepared in the city. It is claimed for the leather of Srinagar that saddles last for ever. Kashmir has a good reputation in furries. The recent law for the protection of game, under which the sale of skins and horns is prohibited, has curtailed the business of furries.

In addition to these, the Kashmir and Jammu artisans have great skill in the manufacturing of copper utensils, shawls, pottery and basket making.

Cement Industry

The raw materials for the manufacture of cement are calcareous and argillaceous materials. These are mixed in suitable proportions to form the raw mix limestone, gypsum, coal, bauxite and clay are the main ingredients of this industry. Limestone and gypsum are available in large quantities in Baramulla and Anantnag districts.

The Wuyan Cement Factory is the largest cement supplier to the Valley of Kashmir. This cement factory was established in 1962 to which raw material is supplied from Uri and Baramulla areas. This factory provides employment to about 275 workers and produces about 2,000 tonnes of cement annually. In 1982, a large cement factory was established at Khrew. This factory is known as the J & K Cements Ltd. Khrew. There are more than 500 workers employed in this factory producing about 600 tonnes of cement a day.

There are several chemicals manufacturing units, tiles factory (Pampore), lignite briquetting plant (Shalateng) glass-making and electric goods manufacturing units in the state.

The Hindustan Machine Tools watch factory was established at Zainakot in the 1970s. There are about 1,500 workers who are employed in this factory. It produces about five lakh watches annually.

The Indian Telephone Industry has been established at Hyderpora (Srinagar). It is a branch of the Indian Telephone Industry, Bangalore which produces telephone parts and accessories. It is a small unit in which about 150 persons are employed.

Industrial Complexes

Since 1980, the Government of Jammu and Kashmir state has been paying adequate attention towards the establishment of agro-based, forest-based and mineral-based industries. The State Industrial Development Corporation has established a number of industrial complexes at Rangreth, Khunamoh and Doabgah in Kashmir and at Bari Brahman in Jammu.

The Rangreth factory assembles television sets, radios, transistors, electronic clocks, stabilizers, electric blankets, tape recorders and jewels for watches. Joinery articles, matches, automobiles batteries and tiles are manufactured at Khonamoh. Bari Brahmana has become an important industrial complex of Jammu, producing detergents, resin products, *vanaspati*, steel rollings, scooters, textiles, sunmica, paper, pistons, hosiery, electric goods, light machines, rubber and plastic goods, chemicals, drugs, paper, printing, and transport goods.

There are assorted cottage industries in the towns and villages of the state. Leather processing, shoe-making, oil pressing, pottery blacksmithy, carpentry, bee-keeping, basket-making, soap-making, fruit preservation etc. are the main cottage and small scale industries in which the rural population in finding employment.

The state of Jammu and Kashmir has not seen much of industrial growth. The industrial underdevelopment is mainly owing to the non-availability of basic minerals and lack of infrastructural facilities. Industrial complexes are however, slowly emerging. The development of new power projects, widening of road-network and coming up of the state on the railway map of the country have helped in creating a conducive atmosphere for the growth and development of industries. A number of industrial areas have been set up and various steps are being taken to help the development of industries. The number of registered factories has gone upto 876 and about 5,600 are the small scale industries in 1998.

10

Tourism

The practice of holiday-making away from one's normal place of residence is known as tourism. Tourists are defined as people visiting a place other than that in which they normally reside, for a period including an overnight stay, for any reason other than following an occupation remunerated in the place visited. This operational definition, therefore, includes certain people travelling for reasons other than holiday-making, (e.g., conference participants, pilgrims) but it is normally impracticable to exclude them when data are collected. The distinction between *Recreation* and *Tourism* is that recreation involves leisure activities of less than 24 hours' duration away from home, whereas tourism involves a longer time scale and therefore, requires more infrastructure in the form of accommodation provision[1]. The International Association of Scientific Experts on Tourism (AIEST) has defined tourism as the sum of phenomena and relationships arising from the travel and stay of non-residents, in so far as they do not lead to permanent residence and are not connected with any earning activity.

The essence of the tourist economy is that the tourist spends, in the chosen holiday area, money that has been earned elsewhere. Charistaller (1964) has pointed out that tourism is a very important free-market agent of the spatial redistribution of wealth[2]. In any economy, tourism acts as an export sector, bringing money into the area, setting up both income multipliers creating jobs[3].

The history of tourism is as old as the human society itself. The early tourism can be traced from the period when man set sail and attempted to know the immediate world around. His inherent zeal for enchanted unknown lands and curiosity for new world culminated into early travels. At the start of the present century, travel and tourism were confined only to affluent few, i.e., rich, religious zealots, conquerors, the well educated and the elites who were fascinated by the enchanting beauty and mysteries of unknown land. Tourism has, however, grown from the pursuits of a privileged few to a mass movement of people, with the urge to discover the unknown places and to seek change in environment and to undergo new experiences[4].

The dramatic transformation of rudimentary tourism into a giant phenomenon has resulted from a set of physico, socio-economic, political, psychological and aesthetic factors. Modern tourism upsurged after the Second World War commencing with a spurt in human mobility and intensive application of mechanization[5]. In the ending part of the 18th century, industrialization resulted into commercialism and urbanization. Consequently, the standard of living of the people went up and there was a concomitant rise in leisure hours compelling people to adopt travel and recreation as an inevitable lifestyle. Also, within a short span, human members increased tremendously particularly in the urban centres. All these factors culminated into ushering a new wave of migration, typically termed as 'Mass Tourism'. At present, touring and outdoor recreation have become a genuine social, economic and psychological need, not for individuals alone but also for our entire society and economy. Tourism development holds immense appeal because of anticipated economic benefits of income and employment[6]. In terms of employment, it is the largest industry that provides jobs to about one in every 16 workers worldwide[7]. It helps in removing the regional disparities, particularly in underdeveloped and backward areas[8].

Kashmir is one of the most beautiful tourist destinations of the world. It used to attract enormous number of domestic and international tourists before 1989. The period between 1989 and 1998 is a lean period from the tourists point of view. The unstable political situation of the state, the militancy and slogan of *Azadi* (independence) discouraged the tourists and recreation-seekers. Tourism, however, is a dominant economic activity in the state. Moreover, about 20 per cent of the workforce of the state is directly or indirectly dependent on tourism.

After agriculture, tourism is the main economic activity in the state. The Jammu and Kashmir state is quite rich in renewable resources (water, forest and fresh air) and scenic beauty. Its lush green forests, rich wildlife, snow-clad peaks, mountainous gorges, giant glaciers, rich fishing grounds, lofty skiing grounds, carpet green turfs, perennial rivers, gushing fountains, floating gardens in Dal Lake, hissing springs, cool breeze, shimmering lakes, invigorating climate, apple-almond orchards, saffron fields and benevolent and kind hearted folk are well known all over the word. In the absence of basic mineral (iron-ore, copper, coal, petroleum, etc.), its productive fields, valuable forests, water and invigorating climate are the great bounties of nature.

Tourism in Kashmir Division

As stated at the outset, Kashmir is one of the most beautiful regions of the world. Characterized with young folded mountains, deep gorges, waterfalls, cascades, scenic beauty and alpine pastures, the state of Jammu and Kashmir provides great opportunities for the domestic and international tourists. In the following paragraphs a brief description of the major places of tourists interest and the games and outdoor activities the tourists can perform have been given.

Around Srinagar

Srinagar, the summer capital of the state of Jammu and Kashmir is situated around the Dal Lake and on both the sides of the Jhelum River. It has many places of interest for the national and international tourists. Having a population of about 7 lakh, it is the primate city in the Valley of Kashmir. It has most of the cultural, educational, religious, recreational and business centres. Apart from the Dal and Nagin lakes the tourists are attracted by the shrines of Hazratbal, Shah Hamdan Mosque, Ziarat (tomb) of Zainul Abidin, Jama-Masjid, Shankaracharya (Takht-e-Sulaiman), Hari Parbat Fort, Panderthan Temple, Harwan, and the Pratap Singh Museum.

Dal Lake

The Dal Lake and its houseboats are one of the main attractions for tourists in Kashmir. The Mughals developed exquisite and elegant gardens like Nishat, Shalimar, Chashma Shahi, Pari Mahal and Naseem Bagh on or around its bank. Some of the Mughal Gardens are excellently maintained even to this day. During the Raj period, Kashmiri rulers did not permit the British (who were as fond of Kashmir's cool climate as the Mughals) to own land here. So they adopted the superbly British solution of building houseboats afloat on Dal Lake. A visit to Kashmir, it is often said is not complete until you have stayed on a houseboat. Subsequently, the Nehru Park and Dal Patti near Hazratbal were developed as the places of attraction for the tourists.

The lake has been divided into three parts (i) Gagribal, (ii) Lokut-dal, and (iii) Bod-dal. Dal Gate, at the city end of the lake, controls the flow of the lake water into the Jhelum River. Within the lake are two islands which are popular picnic spots, i.e., Sona Lanka (Gold Island) and Rupa Lanka (Silver Island). Both are also known as Char Chinar because each one of them has four chinar trees. There is a third island, Nehru Park, at the end of the main stretch of the lake side Boulevard. The water of the lake is, however, highly polluted. There is no sewage disposal system and all the waste from the houseboats goes straight into the lake. The floating fields in the lake and the sediments deposited by the Telbal Nalla are the other pollutants of the lake. The process of eutriphication has accentuated by the application of chemical fertilizers in the paddy fields of the Dachigam-Telbal Nalla.

Despite all these ecological problems, the Dal Lake is a great attraction for the tourists. In fact, the water-dwellers (Hanjis) have a distinct way of life who stay in the water throughout their lives and are buried on the mainland (graveyards) after their deaths. One may observe the mobile shops of vegetables, grocery, mutton, cloths and chemists in the Dal Lake. Thus, a shikara circuit of the lake is a sybaritic experience not to be missed. A leisurely cruise around will take all day, including visit to the Mughal Gardens.

The Boulevard, running alongside Dal Lake, is quite attractive for the tourists. The Shikara Ghats provide access to the houseboats, hotels, restaurant, and shops along the way. Other main road of the city are the Residency Road, connecting the tourist reception centre with the down town area and the Polo View Road lined with handicraft shops and travel agencies[9].

Kashmiris are gifted handicraft-makers and have great dexterity in artifacts. The popular products which attract the tourists are carpets, papier-mache articles, leather and fur goods, wood carving, shawls, scarves, embroidery, honey, cardigans, knitted sweaters and walnut wood articles. Saffron, honey, apples, almond, walnut and dry apricot are also much in demand in the tourist industry[10].

Gulmarg

Gulmarg, literally means 'the meadow of flowers'. It is considered as one of the unparalleled hill resorts in the world. Gulmarg is a wide, green, undulating valley characterized by eskers and terminal moraines, and surrounded by dense fir forests. Situated in the lap of the Pir Panjal Range it is a unique place intersected by the fresh-water channels, dotted over with wood villages. Looking downwards towards the Firozpur Nallah is the vast levelled land devoted to paddy crop in the summer season. On a clear day, Nanga Parbat is visible to the north and Harmukh and Sunset Peak are visible to the south-east.

There are numerous buses running from Srinagar to Gulmarg, many of them of the day tours. The Tourists Hotel, City View Hotel, Tourist Bungalow, the Highway Hotel, Kingsley Hotel, the Green View and Gulmarg Hotel are the reasonable places to stay. Ahdoo's Hotel, next to the bus stand is probably the best place to eat at Gulmarg. A circular road, 11 km in length, runs right round Gulmarg through pleasant pine forest with excellent views over Kashmir Valley[11].

About a six kilometre walk from the Gulmarg bus stop and car park is the important meadow of Khilanmarg. The meadow carpeted with flowers in the spring, is the site for Gulmarg's winter ski runs and offers a fine view of the surrounding peaks and the Kashmir Valley. During the early springs (mid-March to April) as the snow melts, it can be a very muddy hour's climb upto the Khilanmarg meadow.

Gulmarg is famous for winter sports, especially skiing. There is one chair lift, one T-bar and four pomas. All lift operate from 10 a.m. to 5 p.m. A new cable car has been constructed to the top of Mt. Apharwat, and this has opened up some superb new areas above the tree line.

About 13 km from Gulmarg, beyond Khilanmarg at the foot of Apharwat Peak (4511 m) is the location of Alpather Lake. This lake remains frozen until mid-June, and even later in the year one can see lumps of ice-floating in its cold waters. The walk from Gulmarg follows a well-graded pony track over the 3,810 m Apharwat Ridge, separating the lake from Khilanmarg (Fig.10.1).

In addition to these points and places of attraction, there is the *Ziarat* of Baba Rishi. This famous Muslim shrine is on the slopes below Gulmarg and can be reached from either Gulmarg or Tanmarg. The *Ziarat* or tomb, is of a well-known Muslim who died here in 1480 A.D. Before renouncing worldly ways he was a courtier of the Kashmir King Zain-ul-Abidin[12].

Gulmarg is the premier skiing resort. The ski-club of India was established in Gulmarg in 1927. At an altitude of 2,730 m, the slopes of Gulmarg provide runs ranging from below 200 m to 2,000 m and an altitude difference of over 1,500 m, with length of more than 10 km. Moreover, numerous ski-slopes, graded for various levels of skiers, there is a chair-lift of a length of 500 m, providing a ski-run of 700 m. Among the multitude of slopes, there are a few which are served by ski-lift. Most of the skiing becomes centred around these slopes, which are especially suited to beginners and intermediate level skiers. With the construction of Poma Ski-Gondola Cable Car, designed by Polamgalski of France, Gulmarg to Apherwat, the destination offers ski-run for every level including for the most advanced and the most Adventurists (Fig. 10.1).

Gulmarg has the distinction of being Asia's only Heli-skiing Resort Heli-skiing was introduced by Sylvan Saudan in Gulmarg in 1987. In this sport, a helicopter takes the skier upto the Apherwat Peak from where they choose their own descent route. One can mark new tracks almost throughout the entire winter season. There are long chains of mountain valleys and ridges covered with fine powder snow. One can even try a different valley each day with the aid of a helicopter.

Pahalgam

Pahalgam situated at the bank of Liddar River is considered as one of the most beautiful tourist sites in the world. It is about 95 km from Srinagar and

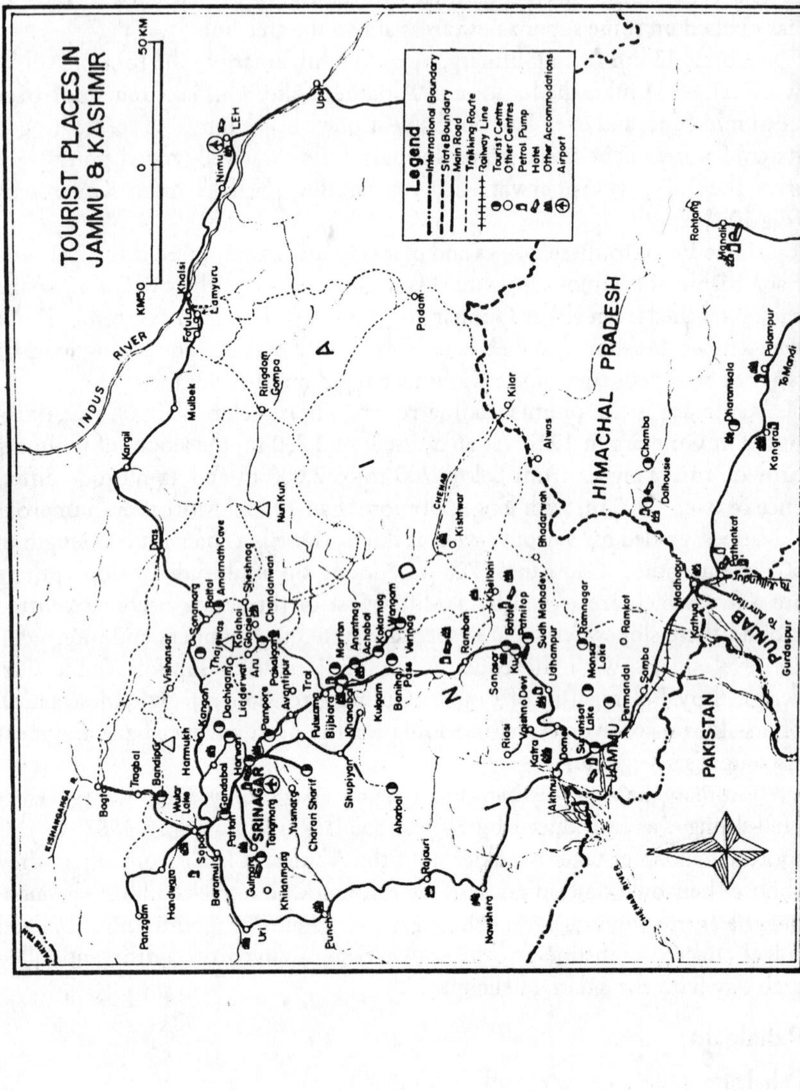
TOURIST PLACES IN JAMMU & KASHMIR
KM50 0 50KM
Legend
International Boundary
State Boundary
Main Road
Trekking Route
Railway Line
Tourist Centre
Other Centres
Petrol Pump
Hotel
Other Accommodations
Airport
INDUS RIVER
SRINAGAR
JAMMU
LEH
HIMACHAL PRADESH
PUNJAB
PAKISTAN
LADAKH
Kargil
Dras
Padam
Nun Kun
Gulmarg
Pahalgam
Anantnag
Kishtwar
Chamba
Dalhousie
Pathankot
Kathua
Udhampur
Rajouri
Punch
Uri
Manali
Rohtang
Dharamsala
Palampur
Kangra

FIGURE 10.1

at 2,130 m above the sea level. The beautiful Liddar River flows right through the town, which is at the junction of the Sheshnag and Liddar rivers and is surrounded by soaring, fir and deodar covered mountains with snow-capped peaks rising behind them (Fig. 10.2).

There are many short walks from Pahalgam and it is an ideal base for longer treks to Kolahoi Glacier or Amarnath Cave. Pahalgam is also famous for many of its shepherds. There are common sight, driving their flocks of sheep along the paths all around the town.

In the vicinity of Pahalgam are the places of tourists interest like Mamaleswara, Baisaran, Aru, and Chandanwari. Mamaleswara is only one km downstream and on the right bank of the Liddar. There is a small Shiva Temple at Mamaleswara. Baisaran is a meadow. There are excellent views over the town and the Liddar Valley from this meadow, five km from Pahalgam. A further 11 km is situated the Tulian Lake at a height of 3,353 m above the sea-level. It is covered with ice for most of the year. Aru is a small village at a distance of about 11 km upstream. It is situated at the Pahalgam-Kolahoi Glacier trek (Fig.10.2). Just outside of Pahalgam on the Amarnath route the *Yog Niketan Ashram* is an interesting place to stay and take *yoga* and meditation course.

Pahalgam, initially a shepherd's village is one of the world's famous resorts and is well known for pony riding, camping sites, trout fish and rarefied clean atmosphere (Fig. 10.1).

Sonmarg

Situated in the Sindh Valley along the Srinagar-Leh, Sonmarg is an excellent base for trekking. Sonmarg means 'meadow of gold' which could derive from the spring flowers or from the strategic trading position it once enjoyed. Sonmarg is situated at the terminal moraines of Thajwas Glacier. Altitudinally, it is 2,740 m above the sea level at a distance of 83 km from Srinagar (Fig.10.2). The tiny village of Baltal is the last settlement in Kashmir, right at the foot of the Zoji La. The Zoji La is the watershed between Kashmir and Ladakh. Thus, Sonmarg area is sheer scenic splendour. The Sindh Valley is surrounded by mountains scaling 9,300 m in height[13].

Sonmarg has a string of shimmering waterfalls and Thajwas Glacier. Gadsar and Satsar lakes are on the periphery of Sonmarg. The Sindh River meanders across the valley and abounds in trouts and Mahaseer—an angler's delight. The other recreational activities here include horse-riding, mountaineering, trekking and fishing. There are tourist huts, a rest house, and some small hotels, restaurants and handicrafts shopping complex.

FIGURE 10.2

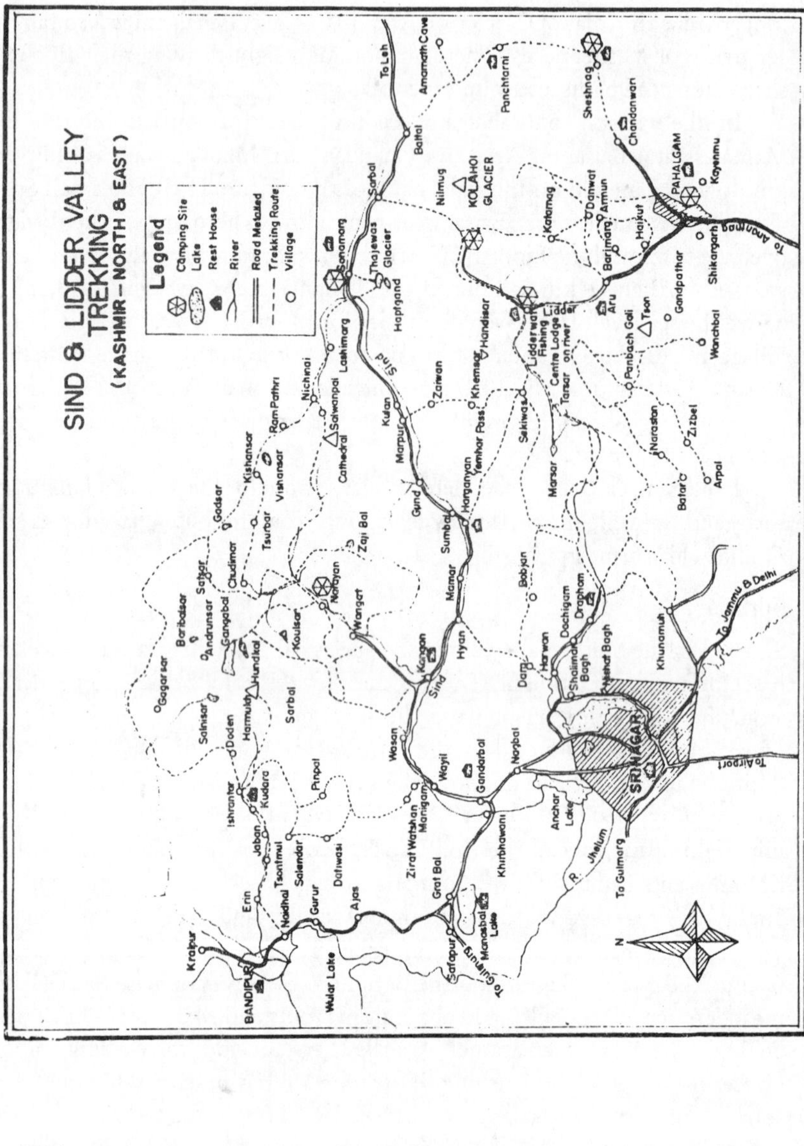
SIND & LIDDER VALLEY TREKKING
(KASHMIR-NORTH & EAST)
Legend
Camping Site
Lake
Rest House
River
Road Metaled
Trekking Route
Village
To Leh
Amarnath Cave
Panchtarni
Sheshnag
Chandanwari
PAHALGAM
Kolahoi Glacier
Sonamarg
Thajewas Glacier
Baltal
Sarbal
Nilmug
SRINAGAR
BANDIPUR
Wular Lake
Manasbal Lake
Anchar Lake
R. Jhelum
To Gulmarg
To Jammu & Delhi
To Airport
To Anantnag
Sind
Lidder
N

Kokernag

Situated at an altitude of 2,020 m, Kokernag is at a distance of about 70 km from the Tourists Reception Centre (TRC) Srinagar (Fig. 10.2). It is famous for the curative properties of springs. It also has a botanical garden with a variety of roses. The Department of Fisheries has collaborated with a Danish firm for large-scale trout breeding at Kokernag. This useful project will annually result in the introduction of fish into the streams. At present, it is being used for the supply of table fish. Kokernag has a Tourist Bungalow, Tourist Huts and a Cafeteria run by the State Tourism Department.

Aharbal

Situated at an altitude of 2,400 m at the debauching point of the Vishav River, Aharbal waterfall (25 m) attracts a large number of picnic-goers during the summer and autumn seasons. It is about 51 km from Srinagar. It has Tourist Bungalow run by the State Tourism Department and a Rest House of the Public Works Department. A charming woodland meadow at Kungwattan is about 5 km ahead. Further ahead at a distance of about 11 km lies the Kousernag Lake which generally remains frozen and covered with icebergs till late June (Fig. 10.2).

Achabal

Situated in the District of Anantnag, Achabal is famous for the Mughal Gardens, which was developed at the spot where a small river comes out gushing out from the limestone topography of the mountain side (Fig.10.1) Achabal is surrounded with villages alive with running water and completely over shadowed by massive chinar trees. The garden has a straight line of fountains and waterfalls and an avenue of chinar on one side. At the head of this beautiful garden is the mountains slope covered with deodar forest. It is a majestic and remarkable sight.

Achabal is about 58 km away from the Srinagar Tourist Reception Centre, situated at an altitude of about 1,667 m. Once the pleasure retreat of Empress Nur Jahan, it has a fine garden, water cascades, cataracts, fine camping ground and a trout hatchery. The Maharaja of Kashmir once pitched a pretty camp here for the entertainment of Lord Minto. Lodging facilities are available in the State Tourism Department, Tourist Bungalow and Tourist Huts.

Yusmarg

Yusmarg situated at an altitude of about 2,377 m is about 40 km from the Tourist Reception Centre, Srinagar (Fig. 10.2). It is a small open vale in the Pir Panjal Range. Giant pines, firs, spruce and deodar cover the mountain slopes on either side of the valley. Pony-riding is the other recreational activity of this silent and tranquil *Marg*. The Nil Nag Lake can be visited from Yusmarg. Boarding and lodging facilities are available in the State Tourism Department Bungalow and the Rest House. There are several Tourist Huts to accommodate the tourists.

Daksum

Situated in the District of Anantnag, it is an enchanting resort griddled by mountains (Fig.10.2). It is about 2,438 m above the sea level at a distance of about 85 km from Srinagar. This resort is more peaceful and less spoilt. From Daksum one can cross to Kishtwar Town of the Doda District (Jammu Division) through the Sinthan Pass.

The places of cultural tourism in Kashmir include Takht-e-Sulaiman (Shankaracharya), in Srinagar, Hari-Parbat Fort, (Srinagar) Pari Mahal, Harwan (once the abode of a famous Buddhist saint), Pandrathan (original City of Srinagar), Awantipora (after Raja Awantivarman who ruled Kashmir from 854 to 883 AD), Martand (this temple is known as the House of Pandus), Verinag (one of the sources of Jhelum), Prihaspora (built by Lahitaditya), and Patan.

The lakes, rivers and streams of Kashmir also provide enormous opportunities for water-skiing white water rafting, canoeing, cruising, ankling, fishing, golfing, cycling, pony-riding, hang-gliding (hot air ballooning) and trekking.

Treks in Kashmir

Being surrounded by the Pir Panjal and the Greater Himalayas, Kashmir provides enormous opportunities for the tourists. There are various treks both within Kashmir and form Kashmir to Ladakh, Zanskar, Kishtwar, Bhaderwah, Rajauri and Poonch. The short Pahalgam to Kolahoi Glacier Trek is particularly most popular and the Pahalgam to Amarnath Cave Trek is well known not only for the natural scenery, but for the great religious festival that takes place annually in the month of July[14].

Pahalgam to Kolahoi Glacier

This short and very popular trek takes only four days from Pahalgam to Pahalgam (Fig.10.2). The first day from Pahalgam takes you to Aru along the bank of Liddar River. The second day's walk takes you to Lidderwat which has a government rest house and a very pleasant campsite, and where the stream from the glacier meets the stream from Tarsar Lake. On the third day you trek up to the lake and back to Lidderwat. The glacier climbing from 3,400 m to 4,000 m, descend from the 5,485 m Kolahoi Mountain. On the fourth day you can either walk straight back to Pahalgam in one day or the trek can be extended another day by walking from Lidderwat to Tarsar Lake and back.

Instead of returning to Pahalgam, three days further trek will take you to Kulan near Sonmarg in the Sindh Valley (a tributary of Jhelum). It is only 16 km from Kulan to Sonmarg which you can walk or travel by bus.

Pahalgam to Amarnath Cave

Amarnath is a sacred religious place of Hindus. The pilgrimage to Amarnath is performed annually in the month of July. At the full noon in the month of July-August, thousands of Hindu pilgrims make the *yatra* to the Shri Amarnath Cave when a natural ice-lingam (stalactite), the symbol of Lord Shiva, reaches its greatest size.

The first days walk out of Pahalgam can be done by jeep and from Amarnath it is possible to continue north to Baltal near Sonamarg, although that is a difficult trek (Fig.10.2).

Sonmarg to Wangat

This 81 km trek takes five days and reaches a maximum altitude of 4,191 m. It starts from Sonmarg (reached by bus from Srinagar), then climbs to Nichinai, crosses a mountain chain and drops down to the pleasant campsite at Kishansar. Another pass has to be crossed on days 3 and 4 when you reach Gangabal Lake. From here it is a steep descent to Wangat, from where you easily bus back to Srinagar (Fig.10.2).

Gangabal Trek

This a 61 km trek which also goes to Gangabal but from the other side. The trek commences from Erin, north of Wular Lake, and takes five days in all. On the fourth day you need rope and ice-access to cross the glacier between the Kundsar and Gangabal lakes. The final day's trek also ends at Wangat. At Naranag, just before Wangat, there is an interesting old temple (Fig.10.2).

Kausarnag Trek

This is a short trek in the south of the Kashmir Valley which ascends into the Pir Panjal mountain. The first day's trek is a short walk, only taking about three hours. With an early start from Srinagar you can bus Aharbal and complete the first walk in the same day. Kausarnag Lake is a beautiful deep-blue stretch of water at 3,700 m (Fig.10.2).

Daksum to Kishtwar

Starting from the south Kashmir Valley at Daksum, this trek is an interesting route to Jammu, although you can also trek from Kishtwar into the Zanskar Valley or to Himachal Pradesh. Daksum is about 100 km from Srinagar and takes about three hours by road. The maximum altitude is reached on the first day's trek to Sinthan Pass. On the last day it is only a short Dadhpath from where buses depart at 10 a.m. and 4 p.m. to Kishtwar.

Pahalgam to Panikhar

This is a hard trek into the Suru Valley which leads to Zanskar. The first two days of the trek follow the Amarnath Cave route. The following days cross the Gulol Gali Pass, climb to the Lonvilad Gali, go over the Chalong Glacier and continue down to Panikhar. From Pannikar you can take the road north to Kargil or east into the Zanskar Valley.

Tourism in Jammu

The Jammu Division of Jammu and Kashmir state is essentially hilly excepting the southern plain along the border of Punjab. The undulating and mountainous topography is drained by the Chenab and Ravi rivers and their tributaries. The incised meanders, gorges, cataracts, river terraces and the snow covered peaks and pine and oak covered slopes present enchanting sights for the tourists. The tourism potential of the state has however, not been adequately developed in the region. It is the cultural tourism in Jammu region, which is attracting more tourists and generating more income for the state.

Jammu with a population of about two lakhs is the summer capital and the second largest town of the state. For most of the tourists, who visit Kashmir, it is the transit point. There are however, several interesting points in the city which fascinate the tourists. Some of the historical and cultural centres which may be visited are as under (Fig.10.3).

Ragunath Temple

The Ragunath Temple is the centre of the city, only a short stroll from tourist reception centre. This large temple complex was built in 1835. The Rambireswar Temple, also centrally located, is dedicated to Lord Shiva and dates from 1883.

Amar Mahal Palace

On the northern outskirts of the town, just off the Srinagar Road, is the Amar Mahal Palace, a curious example of French architecture. The palace museum has a family portrait gallery and another important collections of paintings.

Dogra Art Gallery

The Dogra Art Gallery in the Gandhi Bhavan near the secretariat, has an important collection of miniature paintings, including many from the locally renowned Basholi and Kangra School. The gallery remains open from 7.30 a.m. to 1 p.m. in summer and from 11 a.m. to 5 p.m. in winter but is closed on Monday.

Vaishno Devi

Vaishno Devi is an important pilgrimage centre for the Hindus. It is situated in the Lesser Himalayas near Katra (Fig.10.3). This temple situated in a limestone cave is dedicated to the three mother goddesses of Hinduism. About fifteen lakh (1.5 million) pilgrims visit the Vaishno Devi each year after making a 12 km steep climb from the road head at Katra or taking a shorter and easier climb from a new road. The annual turn out of the Vaishno Devi Trust is about Rs. 2,000 crores. The Trust is so rich that often the State Government of Jammu and Kashmir borrows money from the Vaishno Devi Trust to pay the salaries of its employees.

Patnitop

Patnitop having an altitude of 2,024 m above the sea level is an important hill resort between Kud in the south and Batote in the north. Patnitop is intended to be the nucleus of tourist developments in this area, and there are tourist huts, a Rest House and a Youth Hostel.

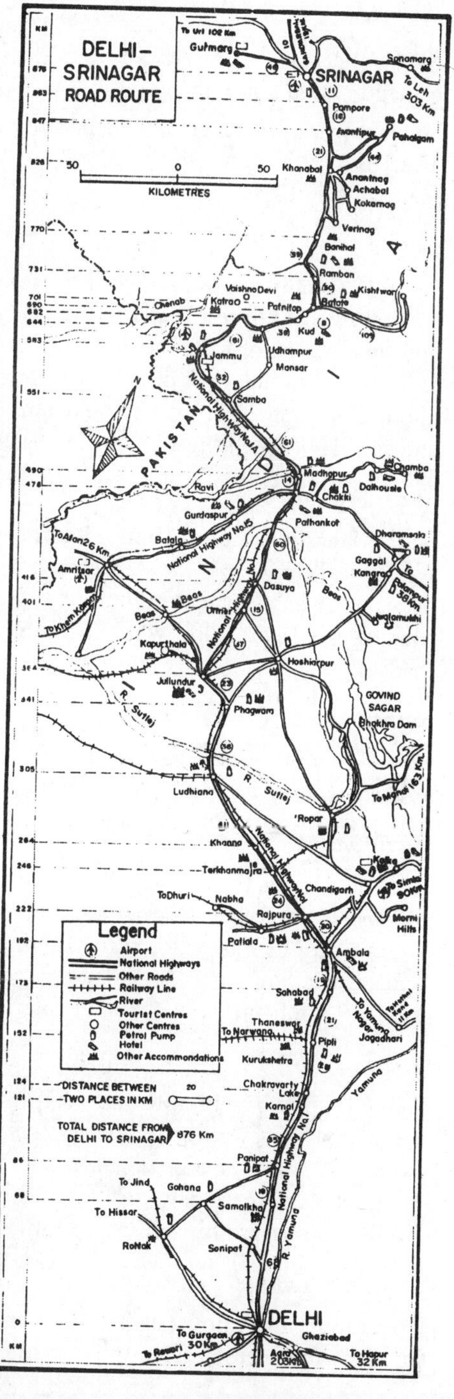

FIGURE 10.3

Kud

Kud is a small township about 6 km south of Patnitop. This is a popular lunch stop on the Jammu to Srinagar route at 1,738 m above the sea level, There is a well-known mountain spring, Swami Ki Bauli about 1.5 km from the road (Fig.10.3).

Batote

Only 12 km further on, and connected to Patnitop and Kud by a number of footpaths, this hill resort at 1,560 m was the overnight stop between Jammu and Srinagar before the Banihal Tunnel was opened. As in Kud, there is a spring close to the town of Batote, known as Amrit Chashma. This spring is about 2.5 km away from Batote. There is a Tourist Bungalow, tourist huts and several private hotels in Batote (Fig.10.3).

Sudh Mahadev

Sudh Mahadev is situated about 8 km to the east of Kud or Patnitop. There is a Shiva Temple at Sudh Mahadev which is visited by a large number of pilgrims on Asadh Purnima (full noon in July-August). About 5 km from Sudh Mahadev is Man Talai, where some archaeological discoveries have been made.

Sunisar and Mansar Lakes

East of Jammu, these lakes are picturesque and the scene of an annual festival at Mansar.

Kishtwar

Well known for the mining of precious stones (agate, ruby, etc.), Kishtwar is situated in the Doda District of the Jammu Division. It is connected with Jammu City by a metalled road. There is a trekking route from Kishtwar to Srinagar via Sinthan Pass. There are many waterfalls around Kishtwar, and about 19 km from the town is the pilgrimage site of Sarthal Devi.

Bhadarwah

It is also known as mini Kashmir. This beautiful valley attracts a large number of tourists every year. Every two years a procession of pilgrims walk from this beautiful high altitude valley to 4,400 m high Kaplash Lake. A week later the three day Mela Patt festival takes place in Bhadarwah. There is a Rest House in this scenic location.

Jawahar Tunnel

This tunnel also known as Banihal Tunnel joins Jammu with Srinagar. During the winter season (December to March) the Valley of Kashmir was often completely cut off from the rest of India before this tunnel was completed in 1967. The 2,225 m long tunnel is 200 km from Jammu and 93 km from Srinagar and has two separate passages. It is rough and damp inside as the water seepages from the roof of the tunnel. As soon as you emerge from the tunnel you are in the green, lush Vale of Kashmir (Fig. 10.3).

Tourism in Ladakh

Ladakh is also known as 'Little Tibet', 'the Moonland' and 'the last Shangir La'. Ladakh is a high altitude plateau to the north of the Himalayas. It is a miniature version of Tibet, the people are Tibetan in their culture and religion, and there are many Tibetan refugees in Ladakh.

The Himalayas are a very effective barrier to rain—few clouds creep across their awesome height and as a result Ladakh is a barren cold desert in which the rate of evaporation is higher than the rate of precipitation. Wherever irrigation water is available the land is highly productive. Plant life is confined to the places where glaciers or melting snow carry water to habitation and alluvial cones. Since Ladakh is barren and dry like the rocks of the moon, it called as the 'Moonland'.

Ladakh could well be a last Shangir La as only in mid-1970s, it was opened to outside visitors. Its strategic isolation is matched by its physical isolation—only from June to September is the road into Ladakh from Kashmir not covered by snow and only since 1979 has there been airline flights into Ladakh. That flight is one of the most spectacular in the world.

Ladakh is a strange worldly place—strange gompas perched on soaring hilltops, shattered-looking landscapes splashed with small but brilliant patches of green, and ancient palaces climbing to sheer rock walls. But most of all, there are the delightful Ladakhis, friendly as only Tibetan people can be and immensely colourful. Ladakh, the highest region in India, comprising deep and narrow valleys between the gigantic mountain ranges of the Himalayas and the Karakoram is one of the India's largest districts. The average height ranges between 3,000 m and 3,650 m above the mean sea level (Fig.10.4). Ladakh's climate is one of sharp contrasts, but it is in this fact itself that its beauty lies. Thousand of feet high, with desolate rocks below and the blue canopy of the sky above, Ladakh has an elusive attraction. A climatic wasteland, it is an aesthetic paradise. For those with a spirit of adventure and enterprise, for those with a love for virgin nature and for those with interest in Tibetology and Buddhism, Ladakh offers an experience that is both rare and transforming[15].

Ladakh comprises of two districts, namely, Leh and Kargil, in the state of Jammu and Kashmir. It is connected with the Kashmir Valley by the Srinagar Leh Road (Fig.10.4). This road winds its way past the Dras Valley

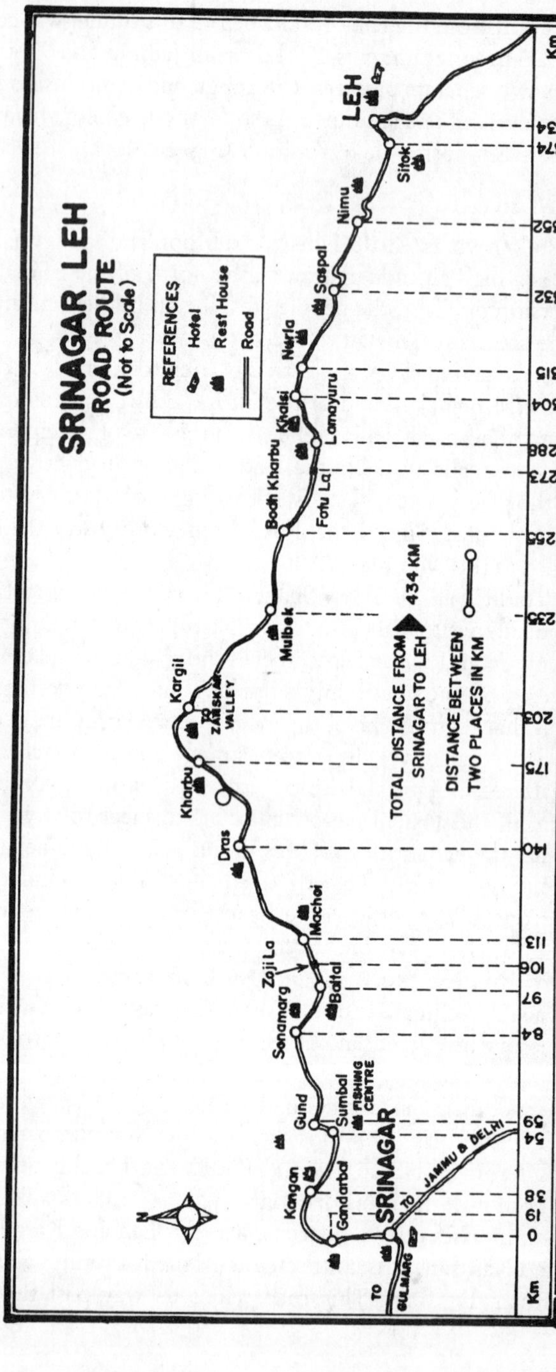
SRINAGAR LEH
ROAD ROUTE
(Not to Scale)
REFERENCES
Hotel
Rest House
Road
N
SRINAGAR
TO GULMARG
TO JAMMU & DELHI
Gandarbal
Kangan
Gund
Sumbal
FISHING CENTRE
Sonamarg
Baltal
Zoji La
Machoi
Dras
Kharbu
Kargil
TO ZANSKAR VALLEY
Mulbek
Bodh Kharbu
Fotu La
Lamayuru
Khalsi
Nurla
Saspal
Nimu
Sitok
LEH
TOTAL DISTANCE FROM SRINAGAR TO LEH 434 KM
DISTANCE BETWEEN TWO PLACES IN KM
Km
0
19
38
54
59
84
97
106
113
140
175
203
235
255
273
288
304
315
332
352
374
434
Km

FIGURE 10.4

Kargil, Ladakh's second largest town with a population of about 40,000, mainly Muslims of the Shia sect.

To the south-east of Kargil lies the largest town of Ladakh—Leh. In the past, Leh was the commercial centre of Central Asia, where numerous languages were spoken and all sorts of goods were traded. Leh, about 3,505 m above the sea level, is a small town of about 1,00,000 people. From here Leh-Manali Road proceeds to Himachal Pradesh.

Srinagar to Leh is about 434 km by road which follows the Indus River for most of the distance. Buses ply along this road daily during the summer season and take two days with two overnight stop at Kargil. Sonamarg is the major town in Kashmir, shortly before you climb up over the Zoji La Pass (3,529 m) and enter the Dras Valley of the Ladakh Division (Fig. 10.4).

Zoji La

This is one of the few unsurfaced stretches on the route. It is also the first pass to snow in winter and the last to be cleared in summer. It is not, however, the highest pass along the route. The other passes get less snow because they are across the Himalayas and in the mountain rainshadow[16]. The road up the pass is breath-taking and there are times when you will wonder if you were sane to make this trip.

Dras

This is the first village after the Zoji La Pass and the place from where road crews clear the road upto the pass for the start of the summer season. In winter, Dras records upto -45°C and is one of the coldest places of India.

Kargil

One of the important trading centres, Kargil is the district headquarters on the way to Leh. The people of Kargil are mainly Muslims and noted for their orthodoxy (Fig. 10.4).

Shergol

Between Kargil and Shergol one crosses the demarcation line between the Muslim and Buddhist areas. The small village of Shergol has tiny gompa perched halfway up the eastern slope of the mountain (Fig. 10.4).

Mulbekh

This is a Buddhist village in which there are two gompas on the hillside above the village. Just beyond Mulbekh is a huge Chamba statue, an image of future Buddha, cut into the rock face beside the road.

Lamayuru

From Mulbekh the road crosses the 3,718 m Namika La, passes through the large military encampment of Bodh Kharbu, the highest pass on the route. Lamayuru is the first of the typical Ladakhi gompas perched on the hilltop with its village of the foot of the hill.

Alchi

Alchi Gompa is unusual in that it is built on lowland, not perched on a hilltop. It is noted for its massive Buddha statues and lavish woodcarvings and artwork.

Lekir and Basgo

Shortly after Saspul a steep road turns off to the Lekir Gompa which also has a monastery school. Close to Leh there is a badly damaged fort at Basgo and the Basgo Gompa with interesting Buddha figures, although its wall paintings have suffered much water damage.

Leh

Situated about 10 km north-east of Indus River at an altitude of 3,540 m, Leh has a population of about 20,000 (1991). Centuries ago it was an important trading centre on the Old Carvan Silk-Route from China. Today, it is merely a military base and fascinating tourist centre.

Overlooking the town, perched on the Tsemo Hill, is the eight-storeyed Leh Palace built by Singe Namgyal in the 16th century. Ladakh is known as the 'Abode of Gods'. It has surreal landscape, meditative monks, and benign people. It is one of the most attractive place for the cultural tourists and pleasure seekers.

Monasteries in Ladakh

Spiritual life in Ladakh centres around monasteries, which served traders and travellers of yore as a place to stay. Each monastery is a conspicuous high building. The path leading to it is usually lined with prayer cylinder called *Chhoskor* which are made of metal and filled with prayer scrolls and charms. Before entering the monasteries a devotee sets the cylinder in motion with a gentle stroke, in the belief that he is sending to heaven prayers equal to the number in the cylinder, multiplied by the number of rotations[17].

Ladakh has about 5,000 lamas and a dozen main monasteries. Most of these can be conveniently visited form Leh as buses ply daily between these monasteries and Leh.

Hemis Gompa

Ladakh's biggest monastery, Hemis Gompa is 49 km form Leh on the Leh-Manali Road. The monastery contains quite a few gold statues and stupas decorated with precious stones. It is a superb collection of thankas including one which is supposed to be the largest in existence and is exhibited only once in 12 years. The next time it can be seen in 2004 A.D. The Hemis festival is held every year either in the second half of June or in the first half of July.

Thinksey Gompa

En route to Hemis Gompa, the Thinksey monastery provides a panoramic view of the green Indus valley from its vantage point atop a hill. It has chambers full of statues, stupas and thankes. There are 100 resident lamas and, allegedly, a nunnery. The Thinksey festival is held annually in early autumn.

Shey Palace and Gompa

About 15 km from Leh towards the Hemis Gompa is the location of Shey Palace of the erstwhile Raja of Leh. Set upon a hill, and housing the largest golden topped victory stupa, the monastry has a two storeyed statue of sitting Buddha in copper and gold which leaves one breathless. Since the monastry is exclusive, it is preferable to make arrangements with the lama before visiting it.

Shankar Gompa

Just 3 km from Leh is the location of Shankar Gompa. This monastry has a formidable collection of miniature statues of pure gold and a number of exciting paintings.

Spituk Gompa

A few kilometres from Leh on the Srinagar-Leh Road, on a hilltop overlooking the Indus, the Spituk Monastry boasts not only some prized thankas, but a chamber with enormous statues of Kali Mata, whose faces are unveiled once a year during the Spituk Festival. The chamber contains an ancient collection of facemasks too. Here the festival is held annually in winter.

Phyang Gompa

The monastry of the red sect of the Buddhists, Phyang Gompa lies 20 km short of Leh on the Srinagar-Leh Road. The monastery possesses some exquisite statues of thankas. Phyang festival is held during summer.

Apart from being the spiritual hub of all socio-economic activities, the monasteries are also the centres of Ladakh's cultural heritage. The religious fervour of the people finds expression in the annual festivals celebrated here. These festivals generally follow the lunar calendar and therefore, their dates vary from year to year. One of the festivals which is famous and celebrated with zest is the Hemis Festival. Hemis as stated above is the biggest monastery of Ladakh and the spiritual centre of Buddhists. People from all over Ladakh assemble here to attend this two-day long festival. The festival is preceded by seven days of prayers by the monks in the monastery. The most impressive part of the festival is the famous mask dance which lamas of the monastery perform continuously for two days. The masks represent different gods and demons. The whole festival signifies the triumph of good over evil. The day marks the birthday of Guru Padmasambhava, the founder of the sect.

Trekking in Ladakh

Ladakh is a region providing enormous opportunities for the trekkers. While going on trekking the trekker should own a tent, a good sleeping bag, sufficient cotton and woolen clothing and some tinned food and juices. Essential medicines, ointments, lotions and creams for sunburns and chaffing would be useful. Use snow-goggles when in snow. Do not eat raw fruit and other eatables at market places. Drink as much liquid as possible and avoid excessive perspiration. It is best to avoid the consumption of alcohol. If you are a heart patient, consult your doctor before going. Respect the religious sanctity of monasteries as these are fountainheads of Buddhist religion and culture. Do not drink alcohol or smoke within a monastery, remove shoes before entering any of its chambers. Do not touch statues. Do not use photflesh inside a gompa as it is likely to damage the colour of the frescoes. Do not take photographs of bridges, strategic locations, military camps and installations as this is strictly prohibited[18]. Some of the important treks are as under.

Dras to Sanku

This is a three-day trek into the Suru Valley, joining the Kargil-Padum Road at Sanku. It is an alternative route to the road down from Kargil.

Kargil to Padum

It is a seven-day trek from Kargil to Srinagar. When all bridges are open the route is accessible from early June to late October. The first dasy's travel is mainly on a surfaced road by bus to Parkchuk, after which one passes through Gulmatango pasture and Rangdum Gompa. Beyond the Rangdum Gompa one has to cross the 4,401 m Pensi La into Zanskar proper. On the last day's walk from Phe to Padum one crosses the river and passes by the Sani Gompa, one of the most important in Zanskar.

Padum to Lamayuru

There are a number of alternatives for this trek from Padum into Ladakh, intersecting the road at Lamayuru, about half way from Kargil to Leh. The trek starts from Padum to Thonde, as on the short trek to the Zangla and Karsha Gompas. On day third, the difficult ascent to the 4,500 m Shing La Pass has to be made, but on day fourth the Nerag La (4,900 m) is even more difficult and local guide is a necessity. From Photosar, a small village, it is only two days' walk to Lamayuru. There is an alternative route form Photosar that takes a day longer. The alternative route from Padum starts out on the opposite side of the Zanskar River and takes you to the Linguist Gompa before joining up with the first route on fifth day.

Padum to Kishtwar

This trek into the southern part of Kashmir is not especially difficult although it crosses the 5,234 m Umasi La, which can only be done in fine weather. One can not use ponies on this trek but must take porters.

On the second day you reach the Zongkhul Gompa, which can also be visited on a short trek from Padum. Day 3 is a long climb and long descent over the snow-covered Umasi La. On day 4 your Zanskri porters will not continue further and you work with many ascents and descents but the road from Kishtwar, already extending to Galar, is gradually being lengthened[19].

Tourism is an age-old activity in the state, especially in the Valley of Kashmir. Innumerable tourists have visited the remote areas like Amarnath, Naranag, Awantipore and hill resorts like Pahalgam, Gulmarg and Bhadarwah from the time immemorial. During the Medieval period, the Mughal monarchs laid beautiful gardens (Nishat, Shalimar, Pari Mahal Chashma Shali, Nasim Bagh, etc.), around the Dal Lake. They utilized the natural springs, encarved cascades, cataracts and canals. They enriched the land with a variety of vegetation, flowers and orchards and revolutionized the handicrafts by encouraging the skilled workers. The traders and travellers for Yarkand (Central Asia) used to take rest in the valley during the Medieval period. A turning point in the history of tourism is marked by the construction of the Jammu-Srinagar National Highway No.1 in 1948, previously known as the Banihal-Cart Road. With the Valley figuring on the air-map of the country in 1948, the international and domestic tourism received a great boost.

At present tourism is an important industry, attracting millions of peoples from all over the world. The tourists who visit the different cultural, religious, aesthetic and recreational places may be classified into: (i) domestic tourists who largely visit Vaishno Devi (Jammu) and Amarnath Caves (Srinagar), most of Indian tourists visit the Valley of Kashmir and their visits are generally confined to the summer and autumn seasons (May to October). The months of September and October in which the holidays of Durga Puja, Dushehra and Deepawali fall, provide a good opportunity to the domestic tourists to enjoy their holidays in the invigorating climate of the Valley of Kashmir and the tranquil settings of Jammu. Contrary to this, the foreigners prefer to visit the Valley of Kashmir in the autumn and winter seasons, i.e., September to January. This is the period when in the hilly tracts of Jammu and Kashmir trekking and skiing are highly enjoyable.

Table 10.1

Jammu and Kashmir: Tourists Coming to Kashmir Valley 1951-1993 (in thousand)

S.No.	*Year*	*Indian*	*Non-Indians*	*Total*
1	1951	9.33	1.25	10.58
2	1955	48.19	2.83	51.02
3	1960	63.37	11.19	74.56
4	1965	35.70	7.43	43.13
5	1970	131.50	11.50	143.00
6	1975	149.15	19.85	169.00
7	1980	548.49	46.03	594.52
8	1985	465.60	30.02	503.62
9	1990	6.10	4.62	10.72
10	1995	8.35	8.35	

Source: Digest of Statistics 1993-94, Jammu and Kashmir Govt. Directorate of Economics and Statistics, Srinagar.

The temporal growth pattern of tourists in the state has been given in Table 10.1. An examination of Table 10.1 shows that in 1951 that about 11,000 tourists visited the Valley of Kashmir, out of which 9,330 were domestic and 1,125 international tourists. With the passage of time the number of tourists grew substantially being 14.3 lakh in 1970. The influx of tourists was unparallel in 1980 when about 60 lakh tourists visited the Valley. Interestingly enough, there were 46,000 foreigner tourists who came to Kashmir in 1980. There however, occurred a slump in the number of tourists after 1989, the period when the militancy in the state increased and the Valley fell into the grip of political turmoil. In 1995, only 8,350 international tourists in came Kashmir. Under the prevailing political instability the domestic tourists did not venture to visit Kashmir. The sense of insecurity discouraged the pleasure seekers and nature enjoyers. In 1998 as per the official information the number of domestic as well as international tourists has increased. It is hoped that more number of tourists will be visiting the Valley in the years to come and this important will flourish again.

International Tourists in Kashmir

The origin of foreign tourists in Kashmir can be traced back to Bernier, most probably the first European to enter the Valley of Kashmir. After him, innumerable number of tourists have visited the destination. In fact, over the period of time Kashmir has changed from a virgin beauty to a managed tourist attraction, The foreigners come in the valley with different purposes and motivations.

The countries and places of origin of the tourists who arrive in the Valley are as widespread as the globe itself.

In the beginning, Europeans formed the major proportion of the tourists who visited Kashmir. With the improvement in means of transportation and social amenities, Kashmir became more accessible and tourists from different parts of the world started pouring in. In fact, Kashmir attracts a substantial number of tourists from Australia, Austria, Belgium, Canada, Denmark, France, Germany, Hong Kong, Italy, Japan, Malaysia, Netherlands, New Zealand, Russia, South Africa, Spain, Switzerland, Sweden, U.K. and U.S.A. The other countries where from Kashmir is receiving tourists are: Afghanistan, Argentina, Bangladesh, China, Czechoslovakia, Egypt, Fiji, Finland, Hungary, Indonesia, Iran, Iraq, Ireland, Israel, Jordan, Kazakistan, Kenya, Labnon, Libya, Mauritus, Myanmar (Burma), Nepal, Norway, Pakistan, Poland, Russia, Saudi Arabia, Singapore, Tajkistan, Tanzania, Thailand, Turkmemistan, Uganda, Uzabaikistan, Venezuela, Yugoslavia, Bosnia, Zambia, etc.

The growth pattern of foreign tourists in the Valley of Kashmir has been given in Table 10.2.

Table 10.2

Jammu and Kashmir: Temporal Increase in Foreign Tourists-1986-1994

Year	*No. of Foreign Tourists*	*Increase/Decrease*
1986	53,118	
1987	59,168	+6,050
1988	59,938	+770
1989	67,776	+7,838
1990	4,627	-63,149
1991	4,886	+259
1992	9,339	+4,453
1993	8,026	-1,313
1994	15,296	+7,270

Source: Dept. of Tourism, Jammu and Kashmir, Srinagar.

An examination of Table 10.2 shows that over 53,000 foreigners visited the Valley of Kashmir in 1986 which rose to 67,800 in 1989. There occurred a phenomenal decrease in the foreign tourists in 1990 in which only 4,600 tourists from abroad arrived in the Valley. The main cause of this tremendous decrease was the *Azadi* Movement, which led to chaos, militancy and insecurity. This downward trend in the arrival of tourists continued upto 1992. In 1992 only 9,300 tourists came to the Valley while this number increased to 15,300 in 1994. After the partition of the subcontinent, it was 1989 when the maximum number of international tourists (67,800) visited the Valley, when 1990 recorded the lowest number of foreigners, being only about 4,600. The substantial decrease in the number of foreign tourists affected the economy of the state adversely and created many economic and occupational problems for the people. In fact, many of the workers who were dependent on tourism have to change their occupations. The business people and the landless workers had to out migrate from the valley.

Table 10.3

Kashmir Valley: Number of Foreign Tourists (countrywise) (in thousand)

S.No.	*Country*	*1971*	*1981*	*1991*	*1993*
1.	U.K.	2.10	6.89	0.57	0.87
2.	West Germany/Germany	NA	4.28	0.48	0.74
3.	Italy	0.36	1.54	0.59	0.61
4.	France	1.45	5.35	0.50	0.43
5.	U.S.A.	5.63	7.44	0.24	0.41
6.	Australia	0.73	2.98	0.33	0.37
7.	Sweden	0.07	0.21	0.13	0.27
8.	Switzerland	0.42	0.93	0.20	0.26
9.	Japan	0.68	1.26	0.19	0.21
10.	Spain	0.03	2.28	0.10	0.20
11.	Belgium	0.14	0.69	0.07	0.19
12.	Malaysia	0.18	0.54	0.04	0.17
13.	Canada	0.43	1.36	0.12	0.15
14.	New Zealand	0.08	0.79	0.10	0.10
15.	Netherlands	0.37	0.82	0.03	0.08
16.	U.S.S.R./Russia	0.03	0.40	0.1	
17.	S. Africa	0.36	1.14	0.03	0.05
18.	Iran	0.07	1.01		
19.	Others	2.11	6.12	1.17	2.91
	Total	15.24	46.03	4.89	8.03

Source: Digest of Statistics, Jammu and Kashmir, 1993-94, Directorate of Economics and Statistics, Srinagar.

The major countries from where the foreigners visit the Kashmir have been given in Table 10.3.

The maximum number of tourists in 1981 came from U.S.A. (7,440), followed by U.K. 6,890. France, Germany, Australia, Spain, Japan and Italy are the other countries from where a sizable number of tourists visit Kashmir.

The bulk of the tourists in Kashmir is domestic. The two cultural and religious sites, i.e., Vaishno Devi and Amarnath Caves attract a large number of tourists form the Hindi Belt, comprising the states of Bihar, Haryana, Himachal Pradesh, Madhya Pradesh, Rajasthan and Uttar Pradesh. The Bengalis, Punjabis, Gujratis, Maharashtrian, etc., visit the Valley of Kashmir mainly during the Durga Puja and Dushehra holidays in the months of September, October and November.

The data available from the Directorate of Tourism, Jammu and Kashmir (given in Table 10.4) show that in 1960-61 only 400 pilgrims visited the Amarnath Cave. This figure rose to 5,400 in 1993-94, recording a continuous upward trend during the last 35 years. Despite political instability and militancy, more and more number of people are visiting Amarnath. It shows a revival of Shivaits Hinduism.

The number of people visiting the Vaishno Devi has been given in Table 10.4.

Table 10.4

Jammu and Kashmir: Tourists Visiting Vaishno Devi and Amarnath Cave, 1950-1993

(in 000)

S.No.	*Year*	*Amarnath Cave*	*Vaishno Devi*
1	1950-51	NA	0.03
2	1955-56	NA	0.06
3	1960-61	0.04	1.60
4	1965-66	0.07	2.30
5	1975-76	0.08	5.81
6	1980-81	0.02	12.13
7	1985-86	0.42	14.85
8	1990-91	0.05	21.69
9	1993-94	0.56	33.69

Source: Directorate Tourism, Jammu and Kashmir.

An analysis of Table 10.4 vividly shows that in 1950-51 Vaishno Devi was an insignificant pilgrimage place attracting only 300 of pilgrims. In 1960-

61 about 1,600 people made *yatra* to Vaishno Devi while this figure reached to 21,690 in 1990-91 and 33,690 in 1993-94. The phenomenal increase in the number of devotees and their pilgrimage to Vaishno Devi show that Hindu fundamentalism is on the increase and it is not the rural folk alone from northern India who perform pilgrimage to Vaishno Devi, the urban educated, rich and elites are increasingly going on *yatra* to this sacred place. The improved condition of staying and resting facilities developed on the way have made this difficult journey, easy to perform. Even the aged people and children may be seen making *yatra* to Vaishno Devi. Consequently, the number of visitors is increasing at a faster pace. More social amenities and infrastructural facilities need to be developed to make the journey easier for the fast growing number of visitors to the shrine.

Tourism is a composite industry. It consists of various segments which can produce a wide range of products and services. Many of these services often not physically measurable. There is also nothing like a standard tourism product. These features of tourism industry make the task of measuring its economic impact a highly complex subject.

The economic benefits of tourism are mainly identified as: (i) Foreign exchange earning, (ii) Income creation, (iii) Employment generation, (iv) Contribution to state revenue, and (v) Regional development. Improved income distribution and multiplier effect, promotion of international economic cooperation, capitalisation of national resources, stimulation of national entrepreneurship, improvement of transport infrastructure are some of the other benefits of tourism. Many of these effects are, however, not directly measurable particularly in view of yawning data-gaps in the tourism sector.

Tourism is also having many social consequences. These include (i) social integration, (ii) widening of social opportunities, (iii) awakening of awareness of national entity, (iv) reviving and strengthening of cultural heritage, (v) strengthening of international understanding, (vi) projection of national image abroad, (vii) spread of awareness of world affairs, etc.

Lack of scientific planning and decision in the promotion of tourism is also having economic and social costs. In the economic front, it may lead to inflation and scarcity of essential items. The social costs are more serious and irrevocable. These include social, and environmental pollution, drug trafficking and flesh trade, corruption of moral and cultural values, etc. Tourism promotion plan of any country has to be, therefore, consistent with its social, economic, ecological and political systems.

Economic Consequences of Tourism

Tourism is often viewed as a new wave of economic opportunity and is largely responsible for the transfer of wealth from one nation to another and redistribution of income, leading to the gross increase in income in the host community of tourism region. The degree of benefit from tourism however, depends largely upon the level of self-sufficiency of the economy of hosting region. In recent years tourism has been considered as a great transformer of economic change. Consequently, tourism has been considered as an industry. Moreover, tourism which is basically a labour intensive industry proves to resist the pressure of economic recession of nations and combat regional imbalances and economic crunch in regional economy.

It is a difficult task to assess the impact of tourism on the economy of the concerned society. Tourism, however, influences the economy in two ways, i.e., (i) the broad consumption items, e.g., transport, accommodation, drinks, and shopping services which have a direct effect, and (ii) the links between the trades directly involved in tourism which supply the tourism trade with goods and services. The full impact of tourism on economy is not just through expenditure on the tourist trade, it embraces all those industries which involve the process to supply goods and services. In the opinion of Smith, each time a tourist makes an expenditure, a ripple of additional spending is sent through the economy which is known as the multiplier effect[20]

It is well known that tourism generates employment. The development of tourism enhances wages, salaries, rent, interest and profits. Since tourism is labour intensive, the greatest proportion of income comes from wages and salaries. In remote tourist sites like Doda, Kishtwar, Bhaderwa, Sonmarg, Daksum, Gulmarg, Pahalgam, Aharbal, Kargil, Leh, Rangdum, Parkchak, Mulbek, etc., it is the only dominant economic activity that yields good income to the people. Taxation on tourist activity also yields income to the state government. Taxes on hotel bills, clubs, wine-shops, and entry tickets to parks and museums, etc., raise additional income to the government.

Tourism provides employment to thousands of people in the state of Jammu and Kashmir who work in hotels, houseboats, guest houses, restaurant dormitories, dak-bungalows, holiday inns and tent houses.

In addition to employment in the tertiary sector, tourism encourages the development of handicrafts and household industries. The state of Jammu and Kashmir, especially the Valley of Kashmir has always enjoyed a universal fame for the exquisite quality and novelty of its arts and crafts. Travellers, poets, and historians have paid glowing tributes to the skill, dexterity and craftmanship of Kashmiris. Industries like carpet-making, fur, leather and embroidery works, papier-mache and woodcarving, willow works, silk manufacturing, pottery-making, etc., employ a large number of handicrafts and household industries in the state of Jammu and Kashmir.

Transport is the lifeline of tourist industry in the state. Rock carving, cutting and building, and rock blasting and landslide clearing activities provide employment to a large number of people. Moreover, circulatory systems like roads, treks, rope-ways and waterways need enormous labour force for their construction and maintenance. Over 50,000 of skilled and unskilled workers derive their livelihood from these activities who would have been rendered unemployed in the state. Tourism discourage outward migration of labour and attract immigrants. This process strengthens the economy of the state.

In addition to job opportunities, tourism accelerates the souvenir making industry. Moreover, a large number of pony-keepers, taxi drivers *shikarawalas*, helpers, tourist guides, shopkeepers, vendors, and fruit sellers directly depend on tourist industry.

Tourism has, thus, opened a new vista for the regional development in the state of Jammu and Kashmir. In the remote tourist sites away from the urban centre a kind of micro-urbanization has to take place as a consequence of tourists demands. The construction of hotels, guest houses, roads, bridges, railway liners, electricity and gas supplies, sanitation, telecommunication network and host of things are developed to provide the required infrastructural facilities to the tourists.

Tourism is a rich source of earning foreign exchange. International tourists generally buy services from the region/state and therefore, pay for invisible.

Adverse Effects of Tourism on Economy

Despite all these merits heavy rush of tourists in the state instigates numerous socio-economic evils. Some of such vices have been briefly given in the following paragraphs.

1. Development of tourism and influx of foreign tourists lead to inflation. Wealthy tourists having a superior purchasing power result into price rise which may be beyond the affordable limits of most of the local population. The cost of living in the areas of attraction for international tourists goes up substantially.
2. Tourism development lead to phenomenal increase in the land prices especially in the areas of attraction for foreign tourists. For example, in the period between 1955 to 1999 the price of land along the boulevard (Srinagar) has gone up by about 1000 times. To be more specific the price of land along the Boulevard in 1955 was about Rs. 5 per sq yard which rose to Rs.10,000 per sq yard in 1998. Similarly, the price of land in Gulmarg and Pahalgam has recorded a stupendous increase being Rs. 1 per sq yard in 1955 and Rs. 500 and Rs. 600 at Gulmarg and Pahalgam

in 1998 respectively.

3. Tourism in the state of Jammu and Kashmir is largely a seasonal activity. Its seasonal character causes seasonal unemployment in winters and shortage of workers during the peak summer season. The hotels and tourists accommodations remain underutilized during the lean season which is a share of the resource. The seasonality of tourism also forces the labourers to outmigrate in search of jobs which puts great stress on the migrants from the state.
4. The promotion of tourism generally creates undue stress on a developing economy. The tourists expect high standards, the provision of which may upset domestic priorities, creating jealousy and discontent.
5. The non-judicious and disproportional use of land and increase in land price convert small farmers into landless labourers. Small farmers dispose off their tiny holdings for more profit. Ultimately, more labour force is generated which leads to unemployment and discontent among the masses.
6. The native population suffers while the tourists enjoy the social amenities. In the development planning of the state there should be more emphasis on the development of basic amenities, like school, hospitals and public transport. Contrary to this, preference is given to the construction of hotels and recreational centres. Consequently, the outsiders enjoy life and the local people suffer.
7. Tourism hinders the establishment and growth of the productive industries. In fact, tourism diverts scare capital away from industries and agriculture.

Social Implications of Tourism

Tourism is a socio-cultural event for the tourists and the hosts. It plays a vital role in social change and brings gradual and radical changes in the cultural values and traditions of the people. The impact is, however, more marked in the lesser developed societies of the developing countries. Tourism often has significant cultural implications (restoration of historical and archaeological monuments), social and political understanding.

According to a study conducted by the UNESCO, the relationship between hosts and guests in the destination areas can be characterized by five main features: (i) they involve transitory relationship. Tourists come to a place or region for a short-period, so an interaction between hosts and guests has little chance to progress beyond casual and superficial level. (ii) there are spatial and temporal constraints to visitor-host interaction. The hospitality business often becomes exploitative to take the advantage of the seasonal character of tourism here. Tourism facilities and services are concentrated in a few locations, (iii) the visitor-resident meetings lack the spontaneity associated with individual schedules. Package tours, planned interactions and scheduled meetings may be arranged but these are only controlled events and finally become commercial in nature, (iv) the meeting and visitors and residents is general an unequal and unbalanced experience. Residents often feel inferior when they compare their situation to a visitor's apparent wealth, (v) while the visitor is on a holiday-making mood enjoying novel experience, for the resident such events and meeting have become routine without any fun. The social consequences of tourism have been briefly described in the following paragraphs.

1. Transformation in Lifestyle

The dwellers of the hilly and mountainous state of Jammu and Kashmir are generally tradition loving, conservative people. They respect their cultural values appreciably. The society has, however, transformed in the areas of tourists' interest. There is a strong preference of the rich people for the luxury and foreign goods. The presence of tourists make the society more liberal and tolerant. Tourism also reduces suspicion and exclusiveness among the hosts and the guests.

2. International Understanding

Travel helps to broaden the mental horizon of man. Tourists in the Jammu and Kashmir state gain an insight into the customs, traditions and way of life of the people, and home the feeling that man is one in human society. It leads to world peace and promotes international understanding. Tourism can be the best remedy to curb insurgence and violence in the state as it draws people and nations closer to each other. It may help to dissipate prejudice and correct misunderstanding.

3. Cultural Exchange

A tourist arrives with treasures of his civilization and history of his country of which he is an ambassador. Travel and tourism is the best way for establishing a special unwritten and unspoken relationship with people of different countries and different times having different modes of life. Thus, tourism promotes cultural exchange and cultural enrichment of those who travel as well those who are the hosts. When a tourist comes in contact with the place he visits and its population, a social exchange takes place. His social background affects the social structure and mode of life of his destination. He in turn is affected by it and many a times carries back home new habits and ways of life. The culture of a destination attracts people to architectures and historical monuments. Moreover, festivals and exhibitions rely heavily on visitors for their audience and attendance. The educational and cultural significance of tourism implicit in seeing people and living a different exciting life which help to tie the chord of eternal human friendship. Tourism is an education of other peoples habits, food, dress and the gamut of life. It, thus, contributes to the mutual knowledge of man and to the development of sense of hospitality. Tourism has made Jammu and Kashmir known to different corners of the world.

4. Tourism is the Repository of Cultural Heritage

Tourism has been increasingly recognized for its role in saving many of the world's cultural treasures. In Jammu and Kashmir, tourism brings good revenue and helps the government in maintaining archaeological sites. For example, Parihaspora, Pandrathan, Martand, Takht-e-Sulaiman, Pari Mahal, Vaishno Devi. Weaving and woodcarving in the ancient style have been rejuvenated.

5. Tourism and Infrastructural Development

Tourism helps in the development of social amenities. All the hill stations and cultural sites are well connected with metalled roads, post, telegraph and telephone services. The power and water supply are also more regular in the places of tourism interest. With the improvement of these social amenities the rural products (fresh fruits and vegetables) have quick access to the markets.

6. Tourism and Change in Housing Design

Tourism has significantly influenced the design of architecture in the state of Jammu and Kashmir. With increase in resort housing, hotels and guest house have been designed and contracted in a planned way. In Srinagar, the new colonies like Raj Bagh, Kursoo, Rawalpora, Shivpora, Shah Faisal Colony, Berin and Hyderpora, and in Jammu, colonies like Gandhi Nagar, Shashtri Nagar, etc., have been constructed on the new patterns with R.C.C. structures. Unfortunately, the growth of slums in Jammu and Srinagar cities are also because of influx of labourers who are pulled in the region by the forces of tourism.

7. Promotion to Local Arts and Crafts

Tourism is one of the significant factors which keeps arts and crafts of the underdeveloped and developing societies alive. In the state of Jammu and Kashmir tourism is a blessing for the local and regional handicrafts, music, woodwork, carpentry, paper mache and shawl-making. The arts and crafts speak of the grammar of Kashmiri society and act as its representative in the outside world. The scenic splendour and aesthetics of Jammu and Kashmir is reflected specimens of art and handicrafts.

Adverse Effect of Tourism

Despite all these merits of tourism, it has many adverse effects on the society. For example, cultural pollution, deterioration in morality and increase in the rate of crimes.

1. The introduction and imitation of foreign ways of life in the developing societies living in isolation and relative isolation bring about change in attitudes, values and social behaviour of the people. Unfortunately, the younger generation is the most susceptible to the pressure of the demonstration effect created by tourists. The youths attempt to imitate the life-style of visiting tourists. Glimpses of copied western culture may be seen in the popularity of jeans, western pop music and rock dance. As the two cultures come into contact, the local values get transformed, ultimately leading to cultural pollution.
2. The second setback occurs in the form of commercialization of art. Tourists seek local artifacts as souvenirs. The heavy demand for typical artifacts leads to bulk production for the poorly crafted works (referred to as 'airport-art').
3. Close proximity and interaction with tourists influence the food habits of those who are associated with tourist industry, e.g., guides, *shikarawalas,* taxi operators, houseboat owners, pony-keepers and people working in hotels and guest houses. Some of these workers have indulged in gambling delinquency and narcotic drug-trafficking. It is generally believed that tourists are largely responsible for increase in liquor consumption, drug-addiction and smoking in the region. The morality has loosened and some of the youngsters are becoming increasingly permissive in their behaviour, reading morally disgusting literature and enjoying obscene films and club life. Consequently, the age-old sound cultural fabric has been threatened and the society feels frustrated. If these negative impacts of tourism are not checked the cultural ethos of the people of Jammu and Kashmir will be damaged beyond reparation.

Impact of Tourism on Ecology

Tourism is essentially a service industry, depending almost entirely on natural ingredients like aesthetic beauty of mountains, hills, lakes, waterbodies, sylvan canopy, flora, fauna and bracing climate. Tourism is however, not a less polluting industry. It has increasingly been recognized as one of the most polluting industries. Some of the influence of tourism on ecology and environment have been concisely presented in the following paragraphs.

1. Damage to the Natural Environment

With the increase in population and urbanization, the number of tourists is increasing year after year. It is leading to physical deterioration of environment. There is an imbalance between tourist demand and physical carrying capacity of the area offering tourist activity. Under such circumstances the expected symbiotic relationship between tourism and conservation breaks-down resulting into ecological crisis.

2. Pollution

Tourism-generated pollution is now a threat in the state of Jammu and Kashmir. Tourism has brought with it overcrowding, congestion, traffic snarls in places like Srinagar, Jammu, Pahalgam, Gulmarg, Sonmarg, Vaishno Devi, etc. The meadows of Gulmarg and Sonmarg have been trampled, the Dal, Wular, Aanchar and Gangabal lakes have been polluted. The waterbodies, the scenic spots, the *margs* (alpine pastures), the natural springs, glaciers, outwash plains and trekking routes have been degraded largely because of the undisciplined and careless behaviour of the tourists.

The Valley of Kashmir has been considered as the powerful magnet for tourists. It caters to the needs of every type of tourists. Its physiographic, orographic and stratigraphic setting is most attractive but ecologically fragile. The unplanned and unregulated tourism has created many environmental problems in the region.

3. Pollution of Waterbodies

The state of Jammu and Kashmir, especially the Valley of Kashmir is characterized with some beautiful lakes. The ecosystems of these lakes in degrading fast. The Dal Lake of Srinagar, well known for its geographical setting and splendid beauty, is in a pathetic condition. Its crystal-clear water has become highly polluted, giving an obnoxious odour. It has become the victim of eutrophication, siltation, encroachment, vegetables cultivation in floating fields, and sewage disposal. It is not only that its water-covered area is shrinking and it is becoming shallower the quality of its water is one of the worst in the Himalayan region. Wrapers, plastic bags, tins, cans, rage, vegetables peeling, empty cigarette cases, garbage and numerous weeds may be seen floating in its water affecting the look of the lake.

The area of Dal Lake which was 22 sq km in 1931 has shrunk to only about 15 sq km in 1999. Out of the present area of the Dal Lake only 11 sq km contains water. In order to accommodate more tourists in their houseboats and houses, more and more *Hanjis* are settling in and around the lake.[21]

In the mid-1970s, with tourism becoming a key industry for Jammu and Kashmir, the pressure on the lake multiplied. There were less than a hundred houseboats on the Dal and Nagin lakes at the turn of the century, but their number grew to a staggering 1,400 by 1981. The 1981 census recorded that hamlets around the lake had a population of 24,500, an increase of over cent per cent since 1973. About 5,000 people were found to be the permanently residents on the houseboats. Although the tourists population dwindled after *Mujahadeen* (terrorist) violence broke out in 1989, the human population on and around the lake registered a steady growth. The Lakes and water Development Authority (LWDA) believes that about 45,000 people are now residents on the houseboats and hamlets, and assuming just about 5,000 tourists live on them through the year, the lake bore the burden of about 50,000 people.

The heavy deforestation in the catchment of Dachigam and Telbal Nallah, which discharge their water into the Dal Lake have accentuated the process of siltation. According to one estimate, the Dal Lake bears an annual silt load of about 80,000 tonnes. Much of the silt is deposited near its northern shore as it is blocked by a *Bund* (embankment) in the lake. The land, thus, formed has been reclaimed for farming and pastoral activity, and this had been reclaimed for farming and pastoral activity, and this had added to the load of organic waste in the lake. The run-off from the fertilized paddy fields and untreated sewage work as excellent manures leading to much faster growth weeds and eutrification in the lake. The stories of Wular, Anchar and Manasbal lakes are almost identical. Tourism development has, thus, damaged the lakes ecosystems making them less useful for the people.

4. Landslides

The heavy vehicular traffic an the National Highway No. 1 (Jammu Srinagar Road) is one of the main causes of landslides between Batote and Ramban (Nashri area), between Ramban and Ramsu (Fagla Tract) and between Banihal and Qazigund. Similarly, the low altitudinal zones Gulmarg, Pahalgam, Sonmarg, Kishtwar, Bhadarwah, Doda, Udhampur, Rajouri and Poonch experience a large number of landslides every year.

Notes

1. Johnston, R.J., 1990, *The Dictionary of Human Geography,* B. Blackwell Ltd. Oxford, U.K.
2. Christaller, W., 1964, Some Considerations of Tourism Location in Europe, Pap. Reg. Sci. Assoc. 12, pp 95-105.

3. Johnston, R.J., op cit p. 492.
4. Robinson, H., 1976, *A Geography of Tourism*, MacDonal and Evans, London, xxi.
5. Murphy, R.E., 1985, *Tourism: A community Approach,* Routledge, New York.
6. *WTO News*, August, 1990.
7. Report prepared by the U.S.A. based Wharton Econometric Forecasting Associates (WEFA).
8. Garrison, 1989, *WTO Projections.*
9. Crowther, G., et. al. 1990, *India: A Travel Survival Kit,* Lonely Planet Pub. Hawthorn, Australia.
10. Ibid p. 239.
11. Ibid.
12. Ibid.
13. Ibid.
14. Ibid.
15. Department of Tourism, Jammu and Kashmir, *Ladakh,* 1986.
16. Ibid.
17. Ibid.
18. Faried, A., 1998, *Tourism in Kashmir Valley and its Impact on Ecology, Economy and Society*, unpublished Ph.D. Thesis, Jamia Millia Islamia, New Delhi.
19. Ibid.
20. Smith, S. 1983, *Recreation Geography,* London.
21. Swami, P., To Save a Lake, *Frontline,* June 5, 1998, pp. 66-70.

11

Urbanization

Urbanization has been defined differently by different geographers and social scientists. The bases of their definitions are either: (i) demographic attributes, or (ii) structural changes, or (iii) behavioural process.

In the demographic sense, urbanization is an increase in the proportion of urban population to the total population over a period of time. As long as urban population to total population increases there is urban growth which is a process of urbanization. On the basis of structural changes, urbanization means more concentration of secondary, tertiary and quaternary activities leading to increased productivity and industrialization. From the behavioural point of view urbanization is a process leading to changes in attitudes, values, characterized with large population, high density and heterogeneity of its inhabitants. Urbanization is a characteristic of economically advancing nations and it is closely linked with industrialization.

The process of urbanization has a long history. It originated during the Neolithic Period (New Stone Age) when man started the domestication of plants and animals around 10,000 years before present (B.P.). The cultivation of plants enabled men to change their economy, to produce food and to live in permanent settlements. Urbanization started in the Nile Valley (Egypt), Mesopotamia, Indus Valley, Hwang Ho Valley and Central America around 6000 B.C. The urban centres of Mahenjodaro and Harappa may be cited as examples of prehistoric urbanization in the Indian subcontinent. In Kashmir, the sites like Naranag (Wangat Valley), Aishmuqam, and Awantipur are the testimony of some of the old urban centres.

Difficulties of Defining 'Urban'

No statistical study of urbanization is possible unless adequate note is taken of the definition of an urban area, or city or town which varies from county to country and from one census to another. In Greenland (Denmark), for example, a place with 300 or more inhabitants is called an urban area, in Canada all the settlements with more than 1000 people are called urban, while in the Republic of Korea, an urban area must have at least 40,000 inhabitants. In Israel, a town, irrespective of its size of population, is a non-agricultural settlement. In Argentina and U.S.A. any settlement above 2,000 and 2,500 respectively is an urban place. Contrary to this is many of the Asian countries like Bangladesh, China, India, Indonesia and Pakistan, villages may be having 10,000 inhabitants, and therefore, different criteria have been adopted to define and classify urban places in these countries.

In India the census definition of 'urban/town' remained more or less the same over the period 1901-51. It was in 1961 that several modifications were introduced to make the definition more satisfying from the statistical point of view. In census of 1991, the urban settlements have been defined on the basis of the following criteria:

1. All places with a municipality, corporation, cantonment-board, or notified town area committee, etc.
2. All other places which satisfy the following criteria:
   (i) a minimum population of 5000;
   (ii) at least 75 per cent of male working population engaged in non-agricultural pursuits; and
   (iii) a density of population of least 400 persons per sq km. In addition, the marginal cases such as project colonies, areas of intensive industrial development, railway colonies, tourist centres, etc., come under the category of 'urban'

In order to make the international data comparable, the United Nations (1958) suggested that the data on urban population should also be presented according to the standard scale. Consequently, many of the countries have further classified their urban settlements on the basis population size into a number of categories. Keeping this in mind, the census of India classified the urban places into the following six categories:

Class I, those having a population of 100,000 or more,
Class II, those having a population between 50,000 and 99.999;
Class III, those having a population between 20,000 and 49,999;
Class IV, those having a population between 10,000 and 19,999;
Class V, those having a population between 5,000 and 9,999, and
Class VI, those having a population below 5,000

According to the 1981 census of India, the total urban population of the state was 12.60 lakh or 21.05 per cent of the total population as against 25.34 per cent for the country as a whole. The projected urban population for 1991 was 23.83 lakh, out of which 6.09 lakh was residing in slums and shanty colonies. In 1981, there were 55 places in the state, identified as urban, The different categories of urban places and their population have been given in Table 11.1.

Table 11.1

Jammu and Kashmir State: Percentage of Urban Population in Various Categories of Towns and Cities, 1981

Class of Towns	*Total Population*	*Percentage*	*No. of Towns/ Cities*	*Names of Towns/Cities*
Class I (100,000 & above)	829363	66.21	2	Srinagar, Jammu
Class II (50,000-99999)	—	—	—	—
Class III (20,000-49999)	148028	11.82	5	Anantnag, Baramulla, Sopore, Kathua, Udhampur.
Class IV (10,000-19,999)	76593	6.11	6	Bari-Brahmana, Bandipore, Poonch, Pampore, Bijbehara, Ranbirsinghpora
Class V (5000-9999)	129875	10.37	18	Shopian, Ganderbal, Leh, Rajauri, Kulgam, Tral, Samba, Arnia, Kishtwar, Charar-e-Sharief, Mattan, Akhnoor, Handwara, Bhadarwah, Rehambal, Parole, Doda.
Class VI (Less than 5000)	68831	5.49	24	Kathua, Bishnah, Reasi, Katra, Ramnagar, Basohli, Achabal, Kargil, Billawar, Kupwara, Nowshera, Vijaypur, Batote, Pahalgam, Thana Mandi, Uri, Kokernag, Ramban, Qazigund, Sunderbani, Banihal, Chenani, Lakhanpora, Gulmarg.
	1252690	100.00	55	

Source: Census of India 1981.

Table 11.1 vividly shows that Srinagar City alone has about 48.37 per cent of the total urban population of Jammu and Kashmir state. Only two cities, i.e., Jammu and Srinagar, the winter and summer capitals, respectively have 66.21 per cent of the urban population. Srinagar City situated in the fertile Valley of Kashmir is like a primate city which is hindering the growth and development of all the other towns in the region.

After Jammu, Anantnag with a population of 33978 (1981) is the third largest town in the state. Interestingly enough, there is no Class II (50,000-99,999) town in the state. Anantnag, Baramulla, Sopore, Kathua and Udhampur are the five towns which fall in the category of Class III, having a population between 22,900 to 33,978 (1981). Nearly 12 per cent of the total

FIGURE 11.1

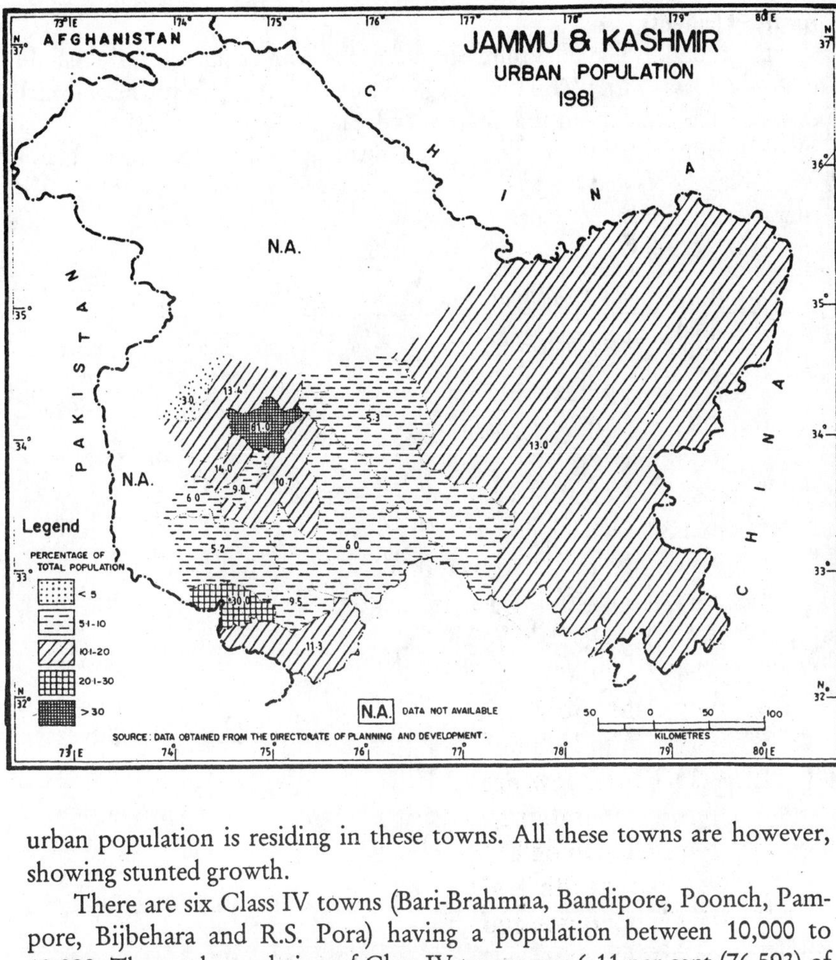

urban population is residing in these towns. All these towns are however, showing stunted growth.

There are six Class IV towns (Bari-Brahmna, Bandipore, Poonch, Pampore, Bijbehara and R.S. Pora) having a population between 10,000 to 19,999. The total population of Class IV towns was 6.11 per cent (76,593) of the total urban population of Jammu and Kashmir state in 1981. The number of Class V towns is 18 (Table 11.1). These towns are inhabited by 10.37 per cent of the total urban population. There are 24 Class VI towns, having a population of less than 5,000 each. These towns are well scattered in the different parts of the Jammu, Kashmir and Ladakh divisions (Table 11.1). There seems to be a lot of migration from these small towns to the bigger cities as a result of which their growth is retarded. Since the role of small towns in the field of industrialization and agricultural and rural development is enormous, their growth is imperative for the balanced urbanization in the state. It is because of this fact that small towns are often called as a 'Bridge' between the urban and the rural universe or between the big cities and the rural settlements.

In general, it may be summarized that the state of Jammu and Kashmir has only Class I cities (Srinagar and Jammu) and rest of the towns are small towns, falling mainly in the Class V and Class VI categories. Many of the small towns are still not having social amenities of drinking water, electricity, good schools and hospitals. In fact, several of the small towns have the characteristics of Indian villages.

Table 11.2

Jammu and Kashmir: Population Variation in Selected Towns and Cities 1971-81

Town/City	*Population in 1971*	*Population in 1981*	*Variation in Population*	*Percentage Increase/ Decrease*
Srinagar	423253	606002	+182749	I 43.18
Jammu	164207	223361	+59154	II 36.02
Anantnag	27643	33978	+6335	III 22.92
Baramulla	26334	33945	+7611	III 28.91
Kathua	17416	23612	+6196	II 35.58
Udhampur	16392	22909	+6517	II 39.75
Bandipora	6201	14218	+8017	I 129.29
Poonch	11981	14171	+2190	IV 18.28
Pampore	8582	10833	+2251	III 26.22
R.S. Pora	6496	10 481	+3985	I 61.35
Leh	5519	8718	+3199	I 57.96
Handwara	5025	6616	+1591	II 31.66
Doda	4395	5396	+1001	III 22.76
Pahalgam	2335*	2626	+291	IV 12.46
Gulmarg	543	511	-32	-5.89

Source: Census of India 1971-81.

In the process of urbanization in Jammu and Kashmir the city life of Srinagar and Jammu appeals to the ordinary middle class or lower middle class people. The social amenities of running water, electricity, educational institutions, medical facilities employment opportunities, and sources of recreation available in these cities are working as the pull factors for the rural migrants.

Looking back over the history of six decades of urbanization in the state of Jammu and Kashmir as revealed in the census reports, it may be said that a number of factors are responsible for the slow growth of most of the urban places and towns. The inclement weather, the mountainous topography, harsh winters, inadequacy of basic minerals (iron-ore, coal), attachment with village life, political instability, slow economic development and the poor accessibility are a few of such factors.

The processes of urbanization and economic change in Jammu and Kashmir state have not been systematically studies. A comprehensive study of the process of urbanization need to be done by an interdisciplinary team, based on historians, geographers, economists, sociologists, demographers, decision-makers, and planners.

Urbanization and Environment

There is a close relationship between urbanization, industrialization and human environment. Environmental change is a continuous process which is taking place right from the primeval stage of human development. With the advent of science, technology, urbanization and industrialization, man is changing the environment with greater vigour. In the last few decades in India, there had been stupendous growth in urban population. Consequently, there is environmental degradation and the quality of life in the urban areas is deteriorating at a faster pace. The problems of increase in greenhouse gases, salination, deforestation, reduction in bio-diversity, water, air, soil and noise pollution are threatening the very survival of mankind.

The experts of environment are of the opinion that owing to fast growth of population, industrialization and urbanization it is not merely the 'quality of life' that is endangered but life itself is endangered by poor drinking water, housing, sanitation, undernutrition, sickness and diseases and natural disasters.

The two largest cities of Jammu and Kashmir state, i.e., Srinagar and Jammu are confronted with the problems of poor housing, slums, shanty colonies, traffic congestion, poor social amenities, air, water and noise pollution. The growth of these cities needs to be checked. In order to improve the quality of life in these cities, efforts should be made to establish and strengthen the network of small and medium-size cities.

The new industrial complexes near Jammu, Srinagar, Khrew (Cement Factory), Sopore, Baramulla, Anantnag, Ramban, deserve special planning efforts to reduce the environmental pollution. In order to achieve these objectives, intensive programme of economic and social improvement should be carried out in the rural areas through balanced agricultural development which will provide increased income to the agriculturist population, permit an effective expansion of social services and include measures to protect the environment.

The stupendous growth of Srinagar City has threatened the Dal Lake which is shrinking at a faster pace. The luxurious growth of weeds has disturbed the lake ecosystem beyond redemption. The untreated sewage discharged into the lake from the surrounding localities has damaged the ecosystem to the extent that it may acquire the shape of a lacustrine plain by the end of the 21st century. The environs of Jammu City are also highly degraded. Deforestation and encroachment of urban settlements into the pastures, agricultural lands, hills and natural areas have induced the process of environmental degradation. A more rational approach for the distribution of population needs to be adopted. Programme should be promoted to make accessible to scattered population the basic social amenities which may go a long way in reducing the outmigration of the rural workforce towards the metropolitan cities of Srinagar and Jammu.

Decentralization of industries is a powerful weapon in directing the settlement pattern in a more orderly fashion. In the state of Jammu and Kashmir, like all the other states of the country, by and large industries have been established in the big cities of Srinagar and Jammu.

The deterioration of environment, both physically and aesthetically, is most apparent in Srinagar and Jammu cities. The dehumanizing effects of life in the slums, ghettos, down towns and old parts of cities, have often been cited as the causes contributing to urban disturbances. Crime rate is usually higher in these neighbourhoods. The environment of eroded localities is less than ideal for human comfort. The people living in the most crowded, smoggy and impersonal cities like the crowded places of Srinagar and Jammu are more vulnerable to mental illness and anti-social behaviour. In the congested parts of Srinagar, Jammu, Anantnag, Baramulla, Sopore, Kathua and Udhampur air pollution is significantly high. Moreover, in the crowded localities, there are inadequate sewage and solid waste disposal systems. Medical care is poor at best, schools are at their worst, public transportation is expensive and inconvenient, crime rates and vandalism are high. This environmental syndrome is reflected by higher death rates among the poor (especially infant and child moralities) than that in the general population.

Srinagar

Srinagar is the largest city in the Western Himalayas. According to the census of 1981, its total population was 586,038. The primacy of the city may be appreciated from the fact that the second largest city of Islamabad (Anantnag) was having a population of only 33,978 in 1981. The projected population in 1998 is more than 7,00,000.

Srinagar City is situated on both the sides of the meandering Jhelum river and around the Dal Lake. The towns of Anantnag and Baramulla are at an equidistance (60 km) from Srinagar. Apart from the summer capital Srinagar is the cultural centre and commercial capital of the region. It had performed religious functions from the establishment of the Panderathan Temple by the king Ashoka in 270 B.C. From the strategic and political point of view it is the nodal point for the entire Valley of Kashmir.

The present Panderathan was the original site of the city which during the period of Ashoka was called Srinagari (the holy city). Subsequently, during the period of Pravasena II, the capital was shifted to the environs of the 'Pradyumnapitha' the present day 'Malkhah Graveyard' near the Hariparbat locality and give it the name of Pravarsenapura with the passage of time the old and the new sites have merged with each other. The city of Srinagar has sprawled on both the sides of the Jhelum river. During the Medieval period with the coming of Rinchan Shah in 1320 A.D. till the end of the Mughal rule (1586-1752), there was a considerable shift in the site of the capital as a result of which they all gave rise to a combination of nuclei and hence we find a vast sprawl of the city. Subsequently, during the period of Afghans, Sikhs and Dogras the city grew moderately, but in the post-independence period the growth of the City of Srinagar is almost unparallel in the hilly regions of India. In addition to administration, business and cultural centre, Srinagar is well known for its tourists attraction all over the world. Millions of tourists flock the city and its environs every year. There is however, a substantial decrease in the number of tourists since 1989. The main reason for the decline of tourist industry is the political unstability.

Jammu

Situated on the undulating southern slopes of Siwaliks on both the sides of the Tawi River, Jammu is the winter capital of the state. With a population of over 3,00,000, Jammu is the second largest city of the state.

Jammu has a twin-city character. The old town lies on a hilltop overlooking the river. The historical buildings—Ram Nath Temple and Maharaja Palace are in this part of the city. Down beside the hill is the railway station, connecting Jammu with New Delhi and other cities of the country.

The Ragunath Temple is in the centre of the city, only a short stroll from the tourists reception centre. This large temple complex was built in 1835. The Dogra Art Gallery, the Amar Mahal Palace are the places of tourists' interest.

The University of Jammu is an emerging institution of great academic importance. There are several government degree colleges, schools and academies in the city. The growth of this city is also stupendous which shall acquire more importance after it is connected with Udhampur by railway line. Near the Jammu Railway Station is the new industrial complex. With the present rate of growth of population, it is hoped that Jammu will become a million city by 2020.

Characteristics of Urbanization

The main features of urbanization in Jammu and Kashmir state are as under:

1. The history of urbanization in the state is quite old as towns and urban places like Panderathan, Naranag, Awantipur flourished during the ancient period.
2. During the ancient and medieval periods, urbanization was associated with the seats of administration, capitals and trading centres.
3. During the period of Sikhs and subsequently after the arrival of English people, the rate of urbanization accelerated.
4. Urbanization in the state is of subsistence nature. It implies that the migrants from rural areas are attracted to the urban centres not for urban environment but for employment. They may be residing in poor houses and may be undernourished but they stick to the cities for employment. This affects adversely the quality of life in the urban places, especially in the cities of Jammu and Srinagar.
5. The cities of Srinagar and Jammu are exploding in their population while rest of all the towns are stagnating. In 1981, about 66.21 per cent of the total urban population was living in the two cities (Srinagar 43.18% and Jammu 36.02%) and 33.79 per cent was in the small towns, having a population of less than 20,000. The town of Gulmarg recorded a negative growth rate in the decade of 1971-81, while Chenani (Distt. Udhampur) was declassified as an urban place.
6. Urbanization in the state of Jammu and Kashmir has a bi-metropolitan apex in which the cities of Srinagar and Jammu dominate, accounting for two-thirds of the total urban population.
7. The urban system in the state is not integrated both spatially and functionally as a result of which there are breaks and imbalances in urban hierarchy. The Class II cities with a population of 50,000 to 99,999 are missing in the hierarchy. At the present rate of urban population growth it is not expected that Anantnag and Baramulla (at present class III towns) will achieve the status of Class II towns (Table 11.2).
8. Towns of Jammu and Kashmir state are growing more on the basis of tertiary rather than the secondary sector.
9. There is not much decentralized urban/industrial development in the state. Most of the district headquarters resemble to the paragon towns of the neighbouring states of Punjab, Haryana and Himachal Pradesh.
10. Urban centres are mostly concentrated in the plain areas. The urban

places in the hilly and mountainous areas are generally small falling mainly in the Class V and Class VI towns with a population of less than 10,000 and 5,000 respectively.

Table 11.3

Jammu and Kashmir State: Towns/Cities-Decadal Variation in Population 1971-81

Variation	*No*	*Names of the Districts*
Above 40 per cent	10	Bandipore, Bijbehara, Jammu, Kargil, Leh, Mattan, Rajauri, R.S. Pora, Srinagar, Uri.
30-40 per cent	9	Bishnah, Handwara, Kathua, Katra, Kishtwar, Kokeranag, Ramban, Samba, Udhampur.
20-30 per cent	11	Ahknoor, Anantnag, Arnia, Baramulla, Doda, Kulgam, Lakhnpore, Pampore, Shopian, Sopore, Tral.
Below 20 per cent	7	Basohli, Batote, Bhadarwah, Nowshera, Pahalgam, Poonch, Ramnagar.
Negative Growth	1	Gulmarg

* Project figure 1991.

Comparable data for the towns of Charar-e-Sharif, Ganderbal, Kathua, Kupwara, Pattan, Parolo, Qazigund, Rehambal and Sunderbani for the period of 1971-71 is not available, while Chenani was declassified in 1961.

Source: Census of India 1971-81.

12

Transport and Trade

Development and expansion of cheap and efficient means of transportation and communication are the prerequisite for speedy and balanced economic growth. The natural resources of a region can be judiciously and suitably utilized only if the means of transportation and communications are adequately developed. The development of these infrastructures, however, largely depends on terrain, topography, geo-climatic conditions and the socio-political milieu of a region.

Transportation involves movement of goods, persons, letters, newspapers as well as non-physical transfer of information through phone, radio, telex, wireless, fax, E-mail and internet. Thus, commodities and information over space. Thus, transport and communication involve the transfer of people, commodities and information over space. This transfer takes place along a line of transportation or communication between two places.

Transportation has always been an important activity of man from the most primitive to the most developed states of the world. It is both result and a cause of a developing society. In modern times, development of means of transportation and communications is an indicator of levels of development.

As stated above, the development and expansion of means of transportation are closely influenced by the terrain, slope, nature of rocks, swamps, deserts and presence of thick forests. Their development and maintenance are also affected by the prevailing temperature, rainfall and snowfall. Such topographical features and climatic conditions stretching across mainlines of trade and travel serve as obstacles to land surface transport systems. Moreover, the historical, socio-economic and political factors also exert great influence on the development of transport and communication facilities. In fact, commercial and industrial growth stimulates invention and innovation in agriculture, mining and manufacturing which tend to increase the demand for transportation. Apart from the socio-economic development of region, transportation plays a vital role at the time of national emergency, like war, famines, droughts, floods, natural hazards and other natural calamities. A sound infrastructure in the form of roads, railways, etc., are, thus, necessary for the existence of a nation and for its balanced and sustainable economic development.

The state of Jammu and Kashmir excepting the Jammu Plain and Jhelum Valley, is mountainous in character. Construction of railway lines through the difficult terrain of the Siwaliks, (Middle Himalayas), Greater Himalayas is a difficult proposition. It is not only a stupendous and challenging task, the maintenance of railway tracks in the fragile rock strata and snowy winters demands prohibitive cost. Nevertheless, the state was brought on the Railway Map of India in 1970 when the City of Jammu-Tawi was connected with Pathankot. The total length of railway lines in the state, at present is about 80 km. Apart from Jammu Tawi, Bari Brahman, Vijainagar, Jammu, Samba, Chagwal, Hiranagar, Chak Dayala, Chhan Arorian, Budhi and Kathua are the small railway stations in the state, on the Jammu-Pathankot Railway line.

Efforts are being made to connect Udhampur Town with Jammu by train. Work on this project has already been started. A number of bridges and culverts have been constructed under this project. The railway line was supposed to be completed by 1992, but owing to financial constraints and socio-political factors the work has been delayed. It is hoped that the gigantic task of construction of the Jammu-Udhampur Railway Line will be completed by 2001 A.D.

The railways have successfully played the role of prime mover to the economy and society of the Jammu Division. The Jammu-Udhampur Railway Line project which is under progress is likely to transform the life and economy of the region.

This 53.2 km long Broad-Gauge Project will provide a rail link between Jammu (the winter capital of Jammu and Kashmir state), and Udhampur which apart from being an important district headquarters of the state is also the headquarters of the Northern Command of the Indian Army.

The construction of this railway line poses a challenge to the Civil Engineering skill of the Indian Railways as these young folded Himalayan mountain ranges present the most difficult terrain. The alignment of this railway line passes through steep, undulating and difficult hilly terrain. The project involved cutting through 212 tunnels and construction of 153 bridges built on tall piers. The high level bridges after being designed and constructed for the first time on Indian Railways by the cantilever method. The longest span of pre-stressed concrete box girder bridge is 102 m, which is a novel feature of the Indian Railways. This project is also credited for having the highest embankment of 42 m. Once constructed, it will be an engineering marvel. The line is proposed to be commissioned upto Udhampur in another two years (2000 A.D.) There will be four new railway stations between Jammu and Udhampur, namely: Bajalta, Sangar, Manwal, and Ramnagar Road.

The new railway line will provide uninterrupted and smooth rail services to the Kashmir Valley and will give fillip to tourism, thereby generating more employment. Industrial development in the region will accelerate as the transportation of goods and material will be easy and economical. The rail link will further strengthen the integration of the people of this state with the rest of the country, and will bring about a qualitative improvement in the life of the people of the region. The main benefits from the Jammu-Udhampur Railway Line will be :

1. It will provide an economical and efficient supplement to the existing transportation and communication infrastructure of the state.
2. It will decongest the road traffic meant to the Valley of Kashmir and the hinterland of Jammu. It will also serve as a stepping stone for the ultimate target of linking the Kashmir Valley with the rest of the Indian Railway System.
3. The new railway line will give a fillip to the industrial development in the region which in turn will generate employment and sources of income for the people.

A blue print of the Udhampur-Srinagar Railway Line is being prepared. Extension of railway line beyond Udhampur is, however, a difficult and expensive project. The steep slopes, the weak rocks formation around *Nashiri*-between Batote and Ramban, the landslides zone of *Fagla* between Ramban and Ramsu, the *Khuni Nala, the Shaitani Nala* and the steep slopes between Banihal and Qazigund are the great hindrances in the development of this railway line. If constructed in the next 20 years, the maintenance of the railway track will be quite expensive. Contrary to this the Valley of Kashmir which is almost a levelled fertile plain deserves the construction of a railway line connecting all the major towns of Kashmir. Very little attention has, however, been paid in this direction, may be because of political and strategic reasons.

Roads

In a state like Jammu and Kashmir over 50 per cent of which is mountainous and almost uninhabited, roads play a vital role in the transportation of people and goods. Throughout the history, the state in general and the Valley of Kashmir in particular have functioned as the focal centre. The Central Asian Trade Route (Silk Route) connecting the subcontinent with Yarkand, khotan, Tashkent, Samarkand, Balkh, Bukhara, Ashikabad, Baku (Azerbaijan),

Armenia and Georgia debauched on to it through the Sindh Valley (a tributary of Jhelum) and the Indus River. The Jhelum river was also an important transport artery.

During the medieval and the modern periods upto 1947, three important routes were followed by traders. Of these the most direct was the road which crossed the Banihal Pass and ran to Jammu; The most popular with the pony men was the old imperial road which ran over the Pir Panjal and reached to railway at Gujarat (Pakistan); and the third was the route known as the Jhelum from Baramulla to Pindi (Pakistan).

After August 1947, the Jammu-Srinagar-Leh Road was given the status of National Highway No. 1-A, the Jammu-Poonch Highway and a number of state roads radiating in all transport system of the state. In other words, the roads provide the most fundamental mode of transport in the state. The metalled roads network connect all the 56 towns and cities (Chapter 11) of the roads connecting, towns, villages, factories, farms, mines and picnic spots which provide the required feeder service. Over 80 per cent of the surface transport in the state is carried on by roads, while in the Kashmir and Ladakh Divisions roads are the only dominant mode of transportation, carrying about 99 per cent of the people and goods.

The National No. 1-A, i.e. Pathankot-Jammu-Srinagar-Uri, and Srinagar-Kargil-Leh, is maintained by the Border Development Board. Since the road in the mountainous areas has been carved out through the weak metamorphic and sedimentary rocks, its maintenance is quite expensive. The tract between Batote and Ramban (*Nashiri*) and between Ramban and Ramsu (*Khuni Nala)* are highly vulnerable to massive landslide. The frequency of landslides on this road is very high during the rainy season and the winter season. In the months of January and February, landslides and heavy snowing block the traffic. It is during this period that thousands of trucks, buses and other vehicles are stranded for weeks together. The *Nashri Bye-pass* (Neera to Batote) provides an alternative route between Batote and Ramban. Similarly, the National Highway No.1 between Srinagar and Kargil remains closed for about five months, i.e., mid-December to mid-May, owing to heavy snow cover at the Zoji La Pass.

At the occurrence of landslides on the National Highway No. 1, the supply of vegetables, cereals, kerosene, and sheep gets disrupted to the Valley of Kashmir. It is during this period when the life of the people of Kashmir becomes difficult. In fact, many of the Kashmiri labourers, artisans and businessmen descend into the cities and towns of Punjab, Haryana, Delhi and Uttar Pradesh to pass their winters.

The Border Roads Organisation has also constructed one road, connecting Manali (H.P.) with Leh (Ladakh). This is one of the highest roads in the world. The average height of the terrain over which this road has been constructed varies between 4,857 to 5,485 m. This road though seasonal in character has reduced the distance between Chandigarh and Leh considerably.

In the state of Jammu and Kashmir, there are a number of state highways. These state highways are the main arteries of trade and commerce and the passengers transport from one region to another. The maintenance of these roads is the state responsibility. The temporal growth of roads and their lengths have been given in Table 12.1.

Table 12.1

Jammu and Kashmir: Road Length Maintained by the State P.W.D.

(Length in Km)

Year	*Road Length*		
	Surfaced	*Unsurfaced*	*Total*
1965-66	3046	1752	4798
1975-76	4974	2341	7315
1985-86	7808	2458	10266
1995-96	9774	2662	12436

Note: Excluding National Highway.

Source: Public Works Department, Jammu and Kashmir.

It may be observed from Table 12.1 that in 1965-66 the total length of metalled road was 3,046 km which rose to 4,974 in 1975-76 and 7,808 in 1985-86. At present the state P.W.D. maintained metalled roads are 9,774 km which has increased almost three times over the last 30 years. Similarly, unmetalled and jeepable roads have been extended into the remote and hilly areas of the state. In 1965-66, the total length of unsurfaced road was 1752 km which rose to 2662 km in 1995-96.

The District-wise length of roads, maintained by the state P.W.D. has been given in Table 12.2.

According to the data available, the total length of roads (metalled and unmetalled) in the year 1995-96 in the state of Jammu and Kashmir was 12,436 km or only 0.89 km per 10 sq km (Table 12.2).

So far as the region-wise availability of road transport is concerned, the Kashmir Division has four km road per 10 sq km area (Fig.12.1). In Jammu and Ladakh divisions this ratio is only 1.67 km and 0.16 km per 10 sq km respectively (Table 12.2).

Table 12.2
Jammu and Kashmir: District-wise Length of P.W.D. Roads

(Length in km)

District/Division	*Area in sq km*	*Road Length*	*One km road per 10 sq. km. area*
Kashmir Division	15948	6480	4.06
Anantnag	3984	1284	3.22
Badgam	1371	810	5.90
Baramulla	4588	1271	2.77
Kupwara	2379	748	3.14
Srinagar	2228	1578	7.08
Pulwams	1398	789	5.64
Jammu Division	26293	4393	1.67
Doda	11691	719	0.62
Jammu	3097	1600	0.51
Kathua	2651	778	2.93
Poonch	1674	178	1.06
Udhampur	4550	682	1.50
Ladakh Division	96701	1563	0.16
Kargil	14036	910	0.64
Leh	82665	653	0.08
Total	138942	12436	0.89

Source: Digest of Statistics, 1995-96.

There are three districts, namely: Srinagar, Badgam and Pulwama in which 5 to 7 km of road is available per ten sq km of (Fig. 12.1). The district of Ladakh with only 0.80 km road per 10 sq km area is the most poorly served district by road. The districts of Doda (0.62 km), Jammu (0.51 km), Kargil (0.64 km) are the other districts in which less than one km road is available per ten sq km of area.

It may be said that despite 50 years of independence, the state of Jammu and Kashmir has a very poor connectivity. There are numerous rural settlements which are not accessible even by the unmetalled roads. The difficult terrain, river gorges and steep slopes are the major hindrances in the construction, development and maintenance of roads in the state.

According to the Directorate of Economics and Statistics the total number of vehicles on roads in 1995-96 was 1,63,586. The public and private transport fleet has been given in Table 12.3.

FIGURE 12.1

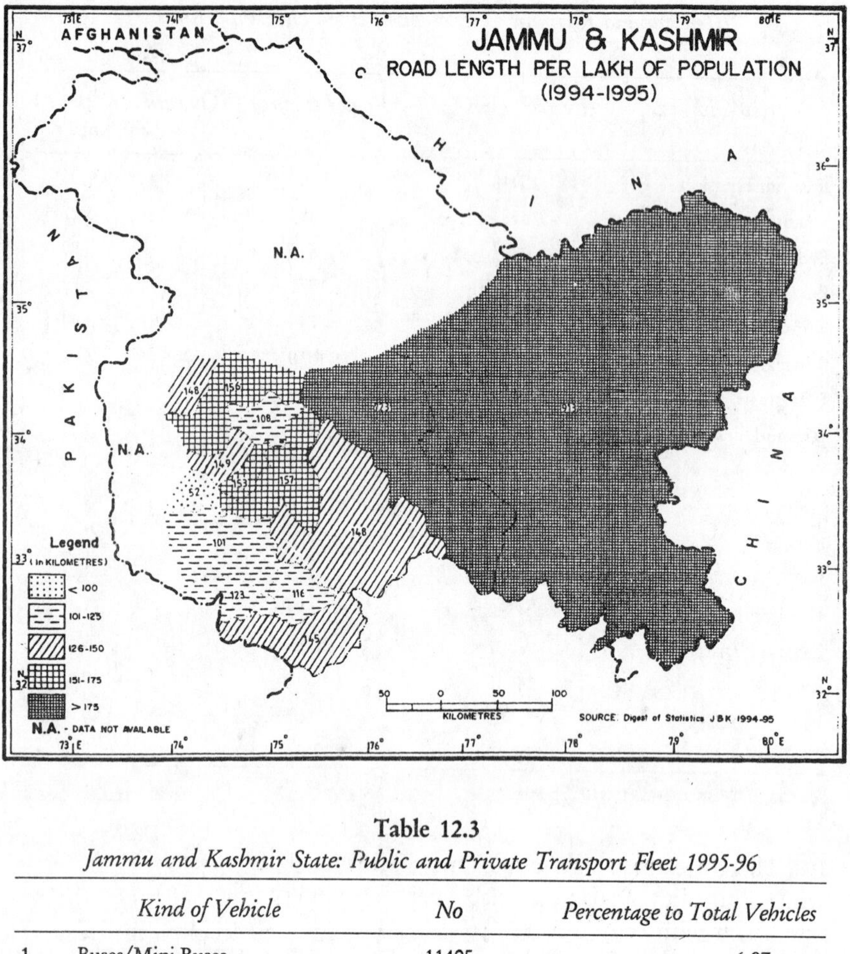

Table 12.3

Jammu and Kashmir State: Public and Private Transport Fleet 1995-96

	Kind of Vehicle	*No*	*Percentage to Total Vehicles*
1.	Buses/Mini Buses	11405	6.97
2.	Trucks	18900	11.56
3.	Taxi	5000	3.06
4.	Private cars	22500	13.75
5.	Jeeps	7500	4.58
6.	Tractors	4700	2.87
7.	Other (Motorcycles, scooters, autorikshaws, trailers, others)	93581	57.21
	Total	163,586	100.00

Source: Digest of Statistic 1995-96.

FIGURE 12.2

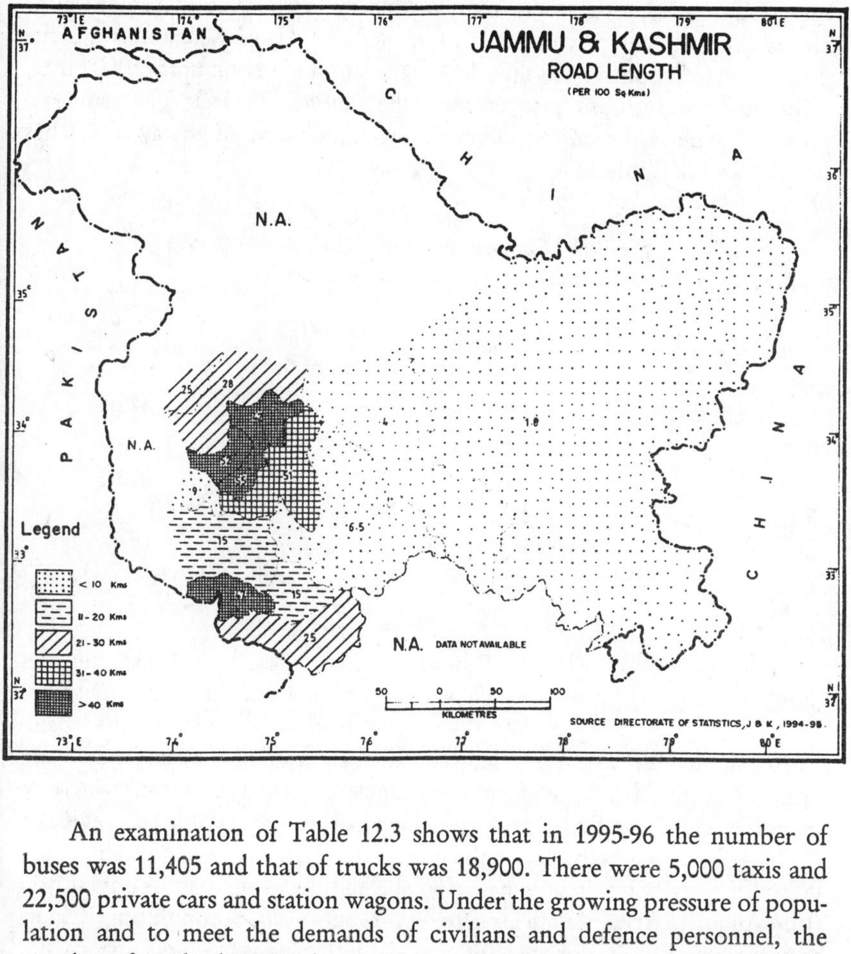

An examination of Table 12.3 shows that in 1995-96 the number of buses was 11,405 and that of trucks was 18,900. There were 5,000 taxis and 22,500 private cars and station wagons. Under the growing pressure of population and to meet the demands of civilians and defence personnel, the number of trucks, buses and private cars is increasing at a faster pace. Lack of railways and very limited scope for water transport are also putting great stress on the road transport.

In the state there are over 11,000 buses and nearly 19,000 trucks. The major burden of transport is shared by these two modes of transportation. There are over 22,000 private cars owned by upper middle and the elite classes. The total number of taxi is 5,000.

The Jammu-Srinagar National Highway No. 1-A, is one of the busiest routes of the country (Fig.10.2). On an average about 800 buses and over 2,000 trucks and army vehicles pass through the Banihal Toll Post every day.

This number varies from summer to winter months. For example, in winters, occasionally, after heavy snowing and rains, the National Highway is closed. During this period the *Nashiri* and the *Khuni Nala* tracts are the worst affected. Contrary to this, during the summer season upto 4,000 buses, mini buses and trucks pass through the Banihal Tunnel. The temporal growth in the traffic on the Jammu Srinagar National Highway No. 1 has been shown in Table 12.4.

Table 12.4

Traffic Passed Through the Banihal Toll Post-1951-1995
(Vehicles Going from Jammu to Srinagar)

Year	*Buses/Trucks*	*Others*	*Total*
1951	16,000	16000	
1961	46,000	2000	48000
1971	78,000	8000	86,000
1981	1,23,000	15000	138,000
1991	1,27,000	15500	142,000
1995	1,32,500	18000	150,000

Source: Jammu and Kashmir Transport Dept., Srinagar, 1996-97.

It may be observed from Table 12.4 that in 1951 only 16,000 vehicles (buses and trucks) entered the Valley of Kashmir from the Jammu Division which rose to 1,32,500 in 1995. The construction of Jawahar-Tunnel (Banihal) facilitated an easy movement of vehicular traffic between Jammu and Srinagar. With the passage of time the vehicular traffic is expected to increase appreciable. To meet the growing demand of the people for the vehicular traffic, it is necessary to expand the National Highway No.1 and to maintain it in a perfect condition even during the snowy winters and rainy season (July-August). An alternative route, i.e., Mughal Road, connecting Rajauri and Poonch with Srinagar via Shopian and Pulwama is to be opened. By doing so the pressure on the National Highway No. 1 will be reduced substantially.

In the hilly and mountainous state of Jammu and Kashmir, the tortuous roads are under great stress. The difficult terrain, sharp bends and steep gradients compound the problems of the drivers. Consequently, about 3,000 road accidents take place every year in which about 800 passengers are killed and a few thousand are injured every year. Unfortunately, the number of accidents is increasing every year. The depleting condition of roads, the heavy traffic flow and the carelessness on the part of driver are the main causes of these mishaps. The government has to implement road safety rules more strictly and have to widen the roads to reduce the hazards of journey, especially on the National Highway No. 1 and on the roads carved out on steep slopes of weak rocks.

Air Transport

In a state like Jammu and Kashmir, characterised by young folded mountains (the Himalayas), gorgeous rivers (Indus, Chenab, etc.), dense forests, and cold desert (Ladakh), development of railways is very difficult and the network of roads can also not be extended beyond certain limits. The remote areas of the state ate still not easily accessible. Under such a hostile geo-climatic environment, air transport can play a very vital role. The importance of airways increases at the time of emergency (war, epidemics, famines, floods, snowing, landslides and avalanches). The development of airways, demands certain threshold of population. The construction of airfields require plain areas for landing and take-off and hazards free weather condition. It is a quick but quite expansive mode of transportation.

Both the capitals of the state (Jammu and Srinagar) and Leh are on the air-map of India. The Indian Airlines connects the airports of Jammu, Srinagar and Leh. There are regular air flights for all these cities. There are several Indian Airlines flights every day between Delhi and Jammu and Delhi and Srinagar. During the peak tourist season (May to July) and in winters when the National Highway No. 1 is blocked owing to snowing and landslides, the Delhi-Srinagar air route becomes awfully busy. The Srinagar Jammu air-route is also a busy route, partly because of the tiring road journey and partly because of the seasonal shift of capital from Srinagar to Jammu in summer and Jammu to Srinagar in winter.

Tremendous progress has been made in the extension, development and maintenance of roads in the state, still much needs to be done to meet the growing need of population and to transport goods in and outside the different parts of the state. Looking at the influx of international tourists, Srinagar deserves the status of an international airport.

Trade

The economy of the state of Jammu and Kashmir is essentially agrarian in character. Most of the workforce of the state is engaged in agriculture, horticulture, sericulture, apiculture, forestry and goats and sheep rearing. Consequently, the state exports a large variety of agro-based and forest-based products. Being poor in basic minerals, the import consists of metallic goods, machinery, iron, chemical products, electric goods, fibres, clothes and apparels. The major items exported from the state have given in Table 12.5.

Table 12.5

Jammu and Kashmir: Goods Exported 1994-95 (in 000 quintals)

	Commodity	*Quantity*
1.	Fresh fruits	7500.00
2.	Dry fruits	250.00
3.	Fibres	40.35
4.	Cotton	13.00
5.	Gums and resins	65.00
6.	Hides and skins	16.00
7.	Chalk and lime	22.00
8.	Glass and glasswares	50.00
9.	Oils	15.00
10.	Wool and woollen goods	10.00
11.	Wood and timber	300.00
12.	Timber manufactured	60.00
13.	Stationery	5.00
14.	Arms and ammunition	50.00
15.	Misc. goods*	438.65
	Total	8835.00

* Misc. goods include-*Anardana*, seeds, bones, drugs, medicines, honey, vegetables, grains, cement, marble, gypsum, paints, varnish, oilseeds, silk goods, embroidery goods, cigarettes, watches, fur and animals.

Source: Digest of Statistics, 1995-96.

Imports

The main commodities of imports in the state of Jammu and Kashmir include machinery, iron steel, metals, electric goods, petroleum, petroleum products, synthetic fibres, rubber, fertilizers, tea, paper, sugar, rice, wheat, oilseeds, groundnut, coal-tar, dyes, crockery (all sorts), drugs and medicines, coal, citrus fruits, vegetables, pulses, leather goods, vehicles, spare motor parts, edible oils, milk powder, stationery, salt, spices, living animals (sheep and goats), military goods, LPG, sports goods, fodder and chemicals. The quantity of goods imported in the state in 1995-96 has been given in Table 12.6.

An examination of Table 12.6 shows that grain, pulses, living animals (sheep, goats, etc.),coal, petroleum, LPG, iron and steal, drugs, milk and milk products, electric goods, chemicals, sugar, books, stationery, chemical fertilizers are the main articles of import. Most of these commodities are imported from the neighbouring states of Punjab, Haryana, Delhi, Rajasthan, Gujarat, and Uttar Pradesh.

Table 12.6

Jammu and Kashmir: Goods Imported, 1995-96

	Commodity	*Quantity in '000 Quintals*
1.	Grains and pules	6500.00
2.	Vegetables	1300.00
3.	Fodder all sorts	700.00
4.	Coal and coke	700.00
5.	Iron all sorts	1250.00
6.	Liquors all sorts	100.00
7.	Vehicles	4000.00
8.	Minerals (cement, marble, etc.)	6125.00
9.	Oils	
	(a.) Edibles oil	230.00
	(b.) Kerosene oil	875.00
	(c.) other oils	35.00
10.	Oilseeds	35.00
11.	Provisions (Ghee, milk, etc.)	450.00
12.	Rubber goods	25.00
13.	Sugar, etc.	1015.00
14.	Living animals (number)	1075495
15.	Misc. goods*	350.00

*Misc. goods include apparel, asphalt, crockery, paints, fibrous goods, glass and glasswares, tea, rubber, soap. woollens, cosmetics, hosiery, spices, chillies, sports goods, timber, music goods, LPG, etc.

Source: Digest of Statistics, Jammu and Kashmir, 1995-96.

It is also vividly clear from Table 12.6 that the people of Jammu and Kashmir in general and that of the Valley of Kashmir are largely dependent on imported food (sheep, goats, wheat, rice, pulses, milk powder, tea, sugar and clothes etc.). The state is not self-reliant even in the case of food and general provisions.

So far as the import of machinery, metallic, electric and chemical goods is concerned, the state is importing bulk of these commodities from the different parts of India.

Though rich in hydel power potential, the supply of electricity is highly irregular in the state. The people of Kashmir combat severe winters and inclement weather with *Kangri* as power supply is not adequate to meet the domestic requirements. The non-availability of power also comes in the way of tourism, especially during the winter season.

Despite all these difficulties, the state has been striving consistently hard for the expansion and diversification agro-based and forest-based exports. In the last ten years tourism could not grow owing to political unstability and environment friendly industries like electronic goods could not be expanded. A number of steps are under consideration to increase export. Assistance and incentives for the establishment of industries have been given to the entrepreneurs. Facilities for training, market research and rationalisation of institutional arrangements and technical services are under progress. Despite all these efforts the quantum of export of finished goods is not showing signs of upward trend. A package programme for industrial development is needed, if export of goods is to be enhanced substantially.

References

Bamzai, P.N.K., 1980, *Kashmir and Central Asia*, Light and Life Pub., Trivendrum.

Jarrte, H.R., 1977, *A Geography of Manufacturing*, Macdonal and Evans, Playmouth .

National Council of Applied Economic Research, 1969, *Techno-Economic Survey*, of Jammu and Kashmir, New Delhi.

Robinson, H. and C.G. Bomford, 1978, *Geography of Transport* Macdonal and Evans, Playmouth.

Sharma, T.C. et. al. 1987, *Economic and Commercial Geography of India*, Vikas Pub., New Delhi.

White, H.P. and M.L. Senior, 1983, *Transport Geography*, Longman, New York.

Appendices

CONVERSION TABLE

Quantity	BRITISH UNITS TO METRIC S.I. UNITS Imperial unit	 Metric equivalent	METRIC S.I. UNITS TO IMPERIAL UNITS Metric unit	 Imperial equivalent
LENGTH	inch	0.0254 m	centimetre	0.394 inches
	foot	0.3048 m	metre	3.281 feet (1.094 yards)
	Yard	0.9144 m	kilometre	0.621 mile
	mile	1.6093 km		
	nautical mile	1.8532 km		
AREA	square inch	645.16 sq.mm	square metre	10,764 sq ft
	square foot	0.0929 sq.m	hectare (10000 sq m)	2.471 acres
	square yard	0.8361 sq.m	square kilometre	0.386 sq miles
	square mile	2.5899 sq.km		
	acre	0.4047 hectares (4047 sq.m)		
VOLUME	cubic inch	16.3871 cu.cm	cubic metre	35.315 cu ft
	cubic foot	0.0283 cu.m	litre (1000 litres = 1 cu m)	0.22 U.K. gallons
	U.K. gallon	4.5 litres		0.264 U.S. gallons
	1000 U.K. gallons	4.5 cu.m		
	U.S. gallon	3.785 litres		
WEIGHT	ounce	28 g (0.028 kg)	gram	0.035 oz
	pound	0.4536 kg	kilogram	2.205 lb
	U.K. ton	1.016 metric tonne (1016.05 kg)	metric ton (1000 kg)	0.9842 U.K. tons
PRESSURE	pound per sq. inch	6894.8 Newtons per sq m	Newtons per sq metre	1.4504×10^4 lb per sq. in
YIELD	bushel (60 1b) per acre	67.253 kg per ha	100 kg per hectare	1,4869 bushels per acre
TEMPER-ATURE	degree Fahrenheit (^oF)	0.555 degree Celsius (^oC)	Degree Celsius (^oC)	1.8^o Fahrenheit (^oF) $t^oC = 9/5t + 32\ ^oF$.
		$t^oF = 5/9(t—32)^oC$		

GEOLOGICAL TIME SCALE

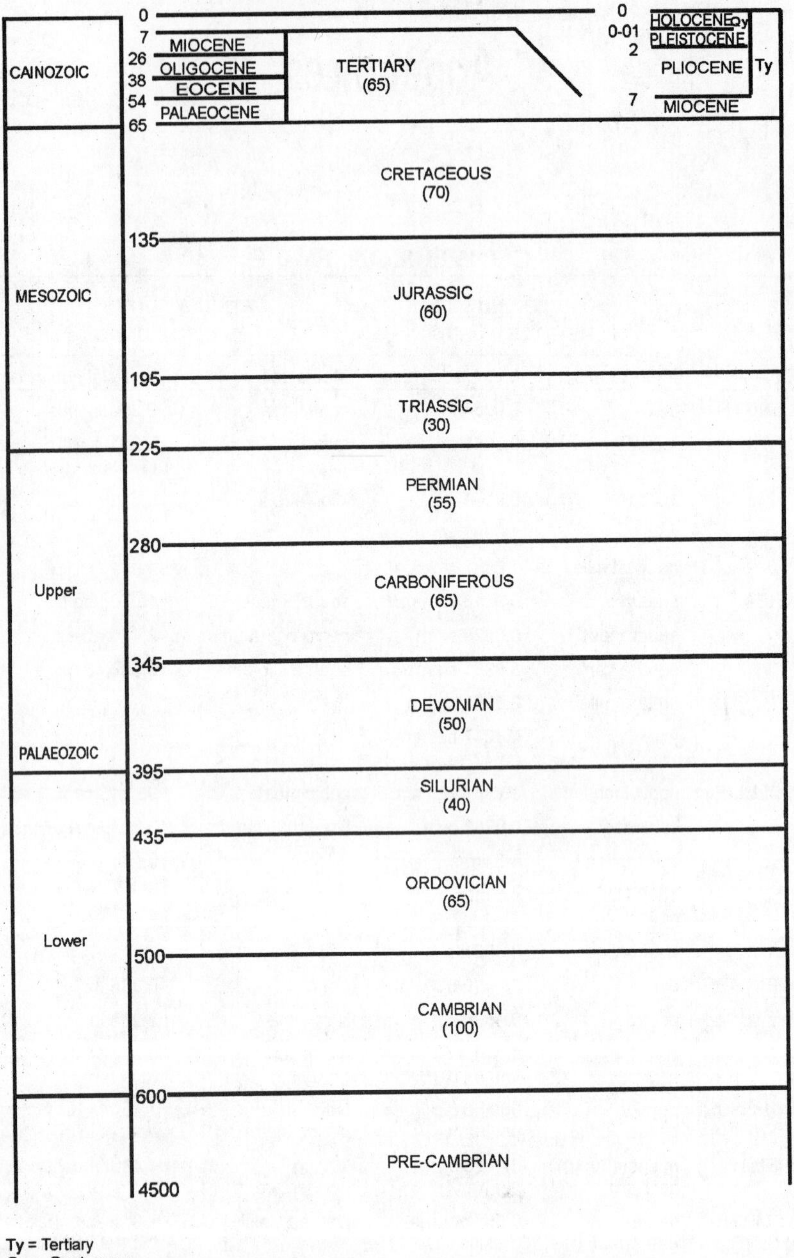

Ty = Tertiary
Qy = Quaternary

The geological time-scale in millions (10^6) of years.

Highest Mountains of the World

	Metres/feet		Metres/feet		Metres/feet		Metres/feet
Everest, Nepal-Tibet	8931.69 M (29,028 F)	Nandia Devi, Uttar pradesh	7891 M (25,645 F)	Muztagh Ata, Sinkiang	7618 M (24,757 F)	Bonete, Argentina	693.7 M (22,546 F)
K2 (Godwin Austen), Jammu and Kashmir-Sinkiang	8692 M (28,250 F)	Rakaposhi, Jammu and Kashmir	7861 M (25,550 F)	Kommunisma, (Pik), U.S.S.R.	7566 M (24,591 F)	Tupungato, Chile	6865 M (22,310 F)
Kangchenjunga, Nepal-Sikkim	8667 M (28,168 F)	Kamet, Uttar Pradesh-Tibet	7830 M (25,447 F)	Pobedy, (Pik), U.S.S.R. Sinkiang	7510 M (24,407 F)	Huascaran, Peru	6832 M (22,205 F)
Makalu, Tibet-Nepal	8555 M (27,805 F)	Namcha Barwa, Tibet	7830 M (25,447 F)	Chomolhari, Bhutan-Tibet	7384 M (23,997 F)	Kailash, Tibet	6778 M (22,028 F)
Dhaulagiri, Nepal	8249 M (26,810 F)	Gurla Mandhata, Tibet	7801 M (25,355 F)	Lenina, (Pik), U.S.S.R.	7202 M (23,406 F)	Yerupaja, Peru	6695 M (21,759 F)
Nanga Parbat, Jammu and Kashmir	8203 M (26,660 F)	Ulugh Muztagh, Tibet	7796 M (25,338 F)	Ancohuma, Bolivia	7081 M (23,012 F)	Sajama, Bolivia	6582 M (21,391 F)
Annapurna, Nepal	8155 M (26,504 F)	Kungur, Sinkiang	7792 M (25325 F)	Aconcagua, Argentina	7026 M (22,834 F)	Illampu, Bolivia	6546 M (21,276 F)
Gasherbrum, Jammu and Kashmir	8145 M (26,470 F)	Tirch Mir, Pakistan	7763 M (25,230 F)	Llullaillaco, Argentina-Chile	6986 M (22,705 F)	Mercedario, Argentina	6526 M (21,211 F)
Gosainthan, Tibet	8090 M (26,291 F)	Amne Machin, China	7692 M (25,000 F)	Illimani, Bolivia	6947 M (22,579 F)	Coropuna, Peru	6486 M (21,079 F)
		Minya Konka, China	7662 M (24,903 F)	Ojos del Salado, Chile-Argentina	6935 M (22,539 F)	Chimborazo, Ecuador	8331 M (20,577 F)

The number of known active volcanoes is 600, of which 109 are submarine. The largest eruption on record is that of Tambora (Indonesia) in 1815; although smaller in terms of total energy released, the most violent eruption on record is that of Krakatoa (Indonesia) in 1883.

Bibliography

Ahmad, M., 1985, 'Spatial Organisation of Tourism in Kashmir State with Special Reference to Sindh Valley', unpublished M.Phil dissertation. University of Kashmir.

Aiyer, N.P.K., 1968, *Field Crops of India,* Bangalore.

Alexander, J.W. and Gibson, L. James, 1979, *Economic Geography,* Englewood Cliffs, N.J.: Prentice Hall.

Bali, U.S., 1982, 'Encroachment of Forests in the Jhelum Valley', Kashmir, Unpublished M.Phil Dissertation, University of Kashmir.

Bamzai, P.N.K., 1980, *Kashmir and Central Asia*, Light and Life Pub. Trivandrum.

Bamzai, P.N.K, 1962, *History of Kashmir,* Delhi.

Bates, C.E., *A Gazetteer of Kashmir.*

Bhan, S.N., 1956, 'Snowfall at Srinagar', *Indian Journal Met. Geophysics,* VII.

Blandford, W.T., *Geology of Western Sind,* Mem. G.S.I. vol. xviii, pt. I, 1879.

Bose, S.C., 1968, *Land and People of Himalayas,* Indian Pub. Calcutta.

Buchanan, R.O., 1974, *An Illustrated Dictionary of Geography,* McGraw Hill, Singapore.

Burrard, S.G., 1933, *A Sketch of the Geography and Geology of the Himalayas and Tibet,* Pt. III Delhi.

Census of India, 1981 and 1991.

Chandna, R.C., 1992, *A Geography of Population*, Kalyani Pub., Ludhiana.

Christaller, W., 1964, 'Some Considerations of Tourism Location in Europe', *Pap.Reg. Sci. Assoc. 12.*

Critchfield, H.J., 1987, *General Climatology,* 4th ed. Prentice Hall, New Delhi.

Crowther, G. et. al. 1990, *India: Travel Survival Kit,* Lonely Planet Pub., Hawthorn, Australia.

Dept. of Tourism, Jammu & Kashmir, Ladakh, 1986.

Dev, J.S., 1983, *Natural Calamities of Jammu and Kashmir*, New Delhi, Ariana Pub., New Delhi.

Diener, Karl, 'Trias of the Himalayas', *Mem. G.S.I.* vol. xxxvi, pt. 3 Digest of Statistics 1912. 1993-94, Govt. of J & K, Srinagar.

Drew, F., 1971, *The Territories of Jammu and Kashmir*, Delhi.

Ehrlich, P.R. et.al. 1977, *Population, Resources and Environment*, Freeman & Co. San Francisco.

Faried, S.A., *Tourism in Kashmir and Its Impact on Ecology, Economy and Society* Unpublished Ph. D. Thesis, Jamia Millia Islamia, New Delhi.

Fazal, A. 1876, *Ain-e-Akbari.*, ed. Blochman, Calcutta.

Gathercoal, E.N., and E.H. Wirth, 1947, *Pharmacognosy*, lea and fiber, Philadelphia. Garrison, 1989, *WTO Projections.*

Gazetteer of Kashmir and Ladakh, 1980, Reprinted in 1974.

Govt. of India, 1970, *Irrigation and Power Projects*, Vivek Pub. Delhi Ministry of Power and New Irrigation, N. Delhi.

Govt. of India, *Irrigation Atlas of India*, 1970, New Delhi.

Govt. of India, *Industrial Indian Manual.*

Govt. of India, Ministry of Tourism, 1987, *Economic Benefits of Tourism*, New Delhi.

Glaeinni, E.A., *Gravity Anomalies and Structure of the Earth Crust*, Survey of India, Dehra Dun, 1932.

Hashia, H., 1988, *Seasonal Migration of Kashmir Labour*, Rima Pub., N. Delhi.

Heim, Arnold, *Himalayan Journal*, ix, 1937.

Holland, T.H. and G.H. Tipper, *Mem. G.S.I.*. xliii, pt I. 1913 and (Second Ed.) *Mem Li-pt. I, 1926, The Dharwar System.*

Husain, M., 1995, *Evolution of Geographical Thought*, 3rd ed. Rawat Pub., Jaipur.

Husain, M., 1986, *Geography of Jammu and Kashmir*, Rajesh Pub., New Delhi.

Husain, M., 1994, *Human Geography*, Rawat Pub. Jaipur

Husain. M., 1996, *Systematic Agricultural Geography*, Rawat Pub., Jaipur.

Husain, M., 1993, *Jaghrafia Alam-e-Islam*, Foundation for Educational Research and Development, New Delhi-25.

Imperial Gazetteer of India, Vol. XV, 1908, London.

Indian Council of Agricultural Research, 1970, *Handbook of Agriculture*, Delhi.

Jarrte, H.R., 1977, *A Geography of Manufacturing*, Macdonal and Evans, Playmouth.

Johnston, R.J., 1986, *Dictionary of Human Geography*, Blackwell, Oxford.

Kachroo, P., 1977, 'A Note on Flora of Kashmir', *Geography of Jammu and Kashmir*, ed. M. Husain, Ariana Pub., New Delhi.

Kendrew, W.G., 1961, *The Climate of Continents,* Oxford University Press, London.

Khatana, R.P., 1985, 'Gujjars and Bakarwals Transhumance in Jammu and Kashmir Himalayas: Responses of Spring Migration', *Geography of Jammu and Kashmir,* ed. M. Husain, New Delhi.

Khatana, R.P., 1992, *Tribal Migration in Himalayan Frontiers,* Vintage Books.

Koul, A.K., 1985, *Home Geography of Jammu and Kashmir,* Srinagar.

Koul, B.L., 1994, 'Shrinkage of Dal Lake', *Sustainable Development and Environmental Impact Assessment,* ed. M. Husain, Rima Pub., New Delhi.

Koul, H.N., 1958, 'A Survey of the Origin of People of Kashmir', *Kashmir Today* vol. II-9.

Kumar, V., 1968, *The Nilmat Purana,* vol. I, Jammu, Lala Rukh Pub. *Race and Trives-Culture of Kashmir,* Srinagar.

Lawrance, W.R., *The Valley of Kashmir,* Navbharat Printing Press, Delhi 1996.

Lidhoo, M.L. , 1985, *Gujjars: Their Attitude Towards Social Changes,* Srinagar.

Mam. A., 1984, The Hanjis of Dal Lake, Unpublished M. Phil. Dissertation, University of Kashmir.

McMahon, 'Microscopic Structure of Some Himalayan Granites', *Rec. G.S.I.* vol. xvii pt. 2, 1984.

Middlecott, H.B., 'Geological Structure and Relations of Himalayas', *Mem. G.S.I.* vol. iii pt. 2, 1864.

Middlemiss, C.S., 'Silurian-Trias Sequence of South-East Kashmir', *Rec. G.S.I.,* vol. x1. pt. 3, 1910

Mir, G.M., 1988, *Impact of Saffron Cultivation on Society and Economy of the Kashmir Valley,* University of Kashmir, Srinagar.

Mohd. A., 1986, 'Regional Structure and Levels of Development of Jammu Division', Unpublished Ph.D. Thesis, University of Kashmir, Srinagar.

Morgan, et. al., 1973, *Human and Economic Geography,* London.

Munshi, A.M., Sidhu, J.S. et. al., 'Improved Cultivation Practices for Saffron', *Indian Farming,* June 1989.

Murphy, R.E., 1985, *Tourism: A Community Approach,* Routledge, New York.

National Council of Applied Economic Research, 1969, *Techno-Economic Survey of Jammu & Kashmir,* New Delhi.

Pareek, H.S., 1976, 'On the Study of Soiled Fuels of Jammu and Kashmir State', *Abstracts, Geol. Survey of India.*

Pattik, J., 1980, *Cultural Heritage of Dogras,* New Delhi.

Ray, S.C., 1970, *Early History and Culture of Kashmir.* Delhi.

Pinfold, E.S., 'Structure and Stratigraphy of N.W. Punjab', *Rec. G.S.I. vol. xlix,* pt. 3, 1918.

Pluyrnin, *Reclamative Soil Science,* New York.

Raza, M. et. al., 1978, *The Valley of Kashmir,* Vikas Pub. House, New Delhi.

Robinson, H. and C.G. Bomford, 1978, *Geography of Transport,* Macdonal and Evans, Playmouth.

Shah, S.A., 1998, 'Growth and Expansion of Saffron Cultivation in Kashmir Division and its Impact on Economy and Society,' Unpublished Ph. D. Thesis, Jamia Millia Islamia, New Delhi.

Shahi, S.S., *The Nomads of Himalayas,* Delhi.

Siddique, M., 1985, 'Urbanization in Jammu and Kahmir: A Regional Analysis,' *Geography of Jammu and Kashmir,* ed. M. Husain, Ariana Pub. New Delhi-12.

Siddique, M., 'Srinagar City in Time and Space', *The Geographer,* vol. xli, No. 1, Jan. 1994.

UNDP, *Human Development Report,* 1995, Oxford University Press, Delhi.

UNDP, *Earth Summit in Focus,* Number 3.

UNFPA, 1997, *Towards Population and Development Goals,* Oxford University Press, New Delhi.

UNICEF, *The Progress of Nations,* 1997.

UNO, Population Division-*1998-99, World Resources: A Guide to The Global Environment,* Oxford University Press, 1998, New York.

UNO, Dept. of Pub. Information, *Environment and Development,* 1999.

Wadia, D.N., 1994, *Geology of India,* 4th ed., Tata McGraw Hill.

Wadia, D.N., 'Cambrian-Trias Sequence of N.W. Kashmir,' *Rec. G.S.I.* vol. 1 xviii, pt. 2, 1934

Wadia, D.N., 'Siwaliks of Postwar and Jammu Hills', *Mem.G.S.I.* vol. 1 pt. 1, 1928.

Wadia, D.N., 'Pleistocene Ice Age Deposits of Kashmir', *Proc. Nat. Inst. Sc. India.*

WEFA Report prepared by the U.S.A. based Wharton Econometrics Forecasting Associates.

White, H.P. and M.L. Senior, 1983, *Transport Geography,* Longman, New York.

Whitten, D.G.A. and J.R.V. Brooks, *The Penguin Dictionary of Geology*, 1977.

Whittow, J., 1984, *The Penguin Dictionary of Physical Geography,* London.

WTO News, August 1990.

Zutshi, B., 1985, 'Population Distribution of Jammu and Kashmir: A Regional Analysis', *Geography of Jammu and Kashmir*, ed. M. Husain, Ariana Pub. Co. New Delhi.

Index